Dynamics of Complex Interacting Systems

Nonlinear Phenomena and Complex Systems

VOLUME 2

The Centre for Nonlinear Physics and Complex Systems (CFNL), Santiago, Chile, and Kluwer Academic Publishers have established this series devoted to nonlinear phenomena and complex systems, which is one of the most fascinating fields of science today, to publish books that cover the essential concepts in this area, as well as the latest developments. As the number of scientists involved in the subject increases continually, so does the number of new questions and results. Nonlinear effects are essential to understand the behaviour of nature, and the methods and ideas introduced to treat them are increasingly used in new applications to a variety of problems ranging from physics to human sciences. Most of the books in this series will be about physical and mathematical aspects of nonlinear science, since these fields report the greatest activity.

The titles published in this series are listed at the end of this volume.

Dynamics of Complex Interacting Systems

Edited by

Eric Goles

Departamento de Ingeniería Matemática,
F.C.F.M.,
Universidad de Chile,
Santiago, Chile

and

Servet Martínez

Departamento de Ingeniería Mathemática,
F.C.F.M.,
Universidad de Chile,
Santiago, Chile

SPRINGER-SCIENCE+BUSINESS MEDIA, B.V.

A C.I.P. Catalogue record for this book is available from the Library of Congress.

ISBN 978-90-481-4734-2 ISBN 978-94-017-1323-8 (eBook)
DOI 10.1007/978-94-017-1323-8

Printed on acid-free paper

Cover design based on a concept by
David Turkieltaub

CONTENTS

FOREWORD vii

MODELS OF SPATIALLY EXTENDED SYSTEMS
Roberto Livi 1

SOME DYNAMICAL PROPERTIES OF ONE-DIMENSIONAL CELLULAR AUTOMATA
Alejandro Maass 35

THERMODYNAMIC FORMALISM OF NEURAL COMPUTATION
Dimitri Petritis 81

LOW COMPLEXITY AND GEOMETRY
Gerard Rauzy 147

FOREWORD

This book contains the courses given at the *Fourth School on Statistical Physics and Cooperative Systems* held at Santiago, Chile, from 12th to 16th December 1994. This School brings together scientists working on subjects related to recent trends in complex systems. Some of these subjects deal with dynamical systems, ergodic theory, cellular automata, symbolic and arithmetic dynamics, spatial systems, large deviation theory and neural networks. Scientists working in these subjects come from several aeras: pure and applied mathematics, non linear physics, biology, computer science, electrical engineering and artificial intelligence.

Each contribution is devoted to one or more of the previous subjects. In most cases they are structured as surveys, presenting at the same time an original point of view about the topic and showing mostly new results.

The expository text of Roberto Livi concerns the study of coupled map lattices (CML) as models of spatially extended dynamical systems. CML is one of the most used tools for the investigation of spatially extended systems. The paper emphasizes rigorous results about the dynamical behavior of one dimensional CML; *i.e.* a uniform real local function defined in the interval [0,1], interacting with its nearest neighbors in a one dimensional lattice.

The paper of Alejandro Maass considers dynamical properties of one-dimensional cellular automata from the point of view of topological dynamics and ergodic theory. The main concepts of these theories are introduced. First, a classification scheme is shown with respect to attractors and equicontinuous points. In particular, the dynamics of positively expansive cellular automata are characterized completely. The symbolic dynamics of the limit sets of cellular automata, or limit complexity, are also discussed.

The survey paper of Dimitri Petritis is devoted to a rigorous study of neural networks. The learning procedure is studied in the rich framework of the thermodynamic formalism. This leads to the analysis of phase transition phenomena, Gibbsian patterns, critical parameters and stochastic dynamics. In this work fundamental concepts are defined, recent results are discussed, and open problems of the theory are pointed out.

The work of Gerard Rauzy deals with the geometric representation of symbolic systems. The concepts relating to language theory and dynamics are introduced. The discussion is focused on the study of symbolic systems of low complexity, which are, in some sense, the more predictable ones. In particular, the geometry of sturmian sequences is fully described.

The editors are grateful to the participants of the School, as well as to the authors of the individual chapters. They are also indebted to the sponsors and supporters whose interest and help was essential for the success of the meeting: Fondecyt, Conicyt, French Cooperation, Fundación Andes, European Community, Centro Física No-Lineal – Chile, French Cooperation – ECOS, Departamento de Relaciones Internacionales, DTI, Departamento de Ingeniería Matemática and CENET of the Universidad de Chile.

The Editors

MODELS OF SPATIALLY EXTENDED SYSTEMS

ROBERTO LIVI
Dipartimento di Fisica, Universitá di Bologna,
and Istituto Nazionale di Fisica Nucleare, Sezione di Bologna,
via Irnerio 46, 40126 Bologna, Italy
&
Istituto Nazionale di Fisica della Materia,
Sezione di Firenze, Largo E. Fermi 2, 50125 Firenze, Italy.
livi@fi.infn.it

ABSTRACT. This series of lectures aims to provide an overview on Coupled Map Lattices as models of spatially extended dynamical systems. Some definitions and generalities about this class of models are reported. Moreover, two specific examples are used for describing their generic properties. In particular, attention is focused on the presence of Chaotic Transients in both periodic and chaotic Coupled Map Lattices, suggesting a close link with complex Cellular Automata evolution rules. The methods used for a quantitative analysis of the dynamical properties have been also presented in some detail. The technical difficulties encountered in a mathematically rigorous description of a seemingly trivial class of models have also been proposed for pedagogical reasons.

1. Introduction

Quite recently the study of Spatially Extended Dynamical Systems (SEDS) has attracted a renewed interest in many disciplines, including mathematics, physics, chemistry, biology, economics, echology and computer science. In fact, beyond usual approaches based on Partial Differential Equations (PDE), new models have been introduced to account for non equilibrium complex evolution observed in fluid convection, reaction diffusion processes, pattern formation and turbulent bahaviours. The most recent class of models is the one of Coupled Map Lattices (CML), which are the subject of these lectures.

CML were introduced just one decade ago by various authors independently [15]. From that time on, they have become one of the most popular tools for the investigation of SEDS. Their success is mainly due to the great flexibility in modelling and to the easy implementation on computers. On the other hand, these features have favoured a sort of *phenomenological* attitude, yielding the classification of a large variety of interesting and, sometimes, peculiar properties that deserve a theoretical interpretation. Some successful efforts have been made in this direction, although the gap between the multi-faceted zoology of CML and a robust theoretical framework is still far from being filled.

The aim of these lectures is to present some of the crucial aspects of CML combining informations obtained from numerical analysis and theoretical , if not mathematical, approaches. In practice, I shall try to single out by specific examples some of the generic properties that characterize CML and that, in my opinion, can be of great help in identifying the basic elements of a theory.

In Section 2 I shall introduce definitions of CML and briefly discuss some general properties. Section 3 will be devoted to describe generic features of CML dynamics emerging in two specific classes: periodic coupled maps (PCML) and chaotic coupled maps (CCML). In Section 4 I shall present some typical tools that can be used to obtain a quantitative description of the different dynamical regimes emerging in CML. Finally, in Section 5 I will discuss, for pedagogical purposes, some rigorous results obtained for superstable PCML.

2. Definitions and Examples

A CML is a deterministic dynamical system evolving by discrete time steps equipped with a spatial structure usually defined on a lattice Λ. On each node $l \in \Lambda$ there is a state variable x_l belonging to a local state space X_l with an uncountable number of elements (in the case where X_l has a countable number of elements CML reduce to Cellular Automata). The state space of the CML, $A = \bigotimes_l X_l$ is the direct product of all the local state spaces, so that any state variable $\mathbf{x} \in A$ is represented by the whole set of local state variables: $\mathbf{x} = \{x_l\}_{l\in\Lambda}$. The CML can be viewed as an updating rule Φ of A into itself, which preserves the lattice structure

$$\Phi\mathbf{x} = \{\Phi_l\mathbf{x}\}_{l\in\Lambda} \tag{2.1}$$

with $\Phi_l : A \to X_l$. In principle Φ_l may act on the whole state space A, with no limitation to any finite subset of A. The most frequently used definition of CML amounts to split the mapping Φ into two distinct mappings F and D, so that $\Phi = D \circ F$. Here F represents a mapping independently acting on each local state variable, i.e. $(F\mathbf{x})_l = f_l(x_l)$, such that $f_l : X_l \to X_l$. At variance, D represents the interaction term $(D\mathbf{x})_l = d_l(\mathbf{x})$. It is worth mentioning that the alternative definition $\Phi = F \circ D$ is also possible, although, in general, it cannot be considered equivalent to the previous one. Nevertheless, in some models these two definitions result to be equivalent from any practical point of view.

One of the most studied versions of CML is the nearest-neighbour coupling case in one space dimension, where f_l is some non-linear map of the interval [0,1] into itself. Using the previous definitions we specialize to the case where $\Lambda = Z$, $X_l = [0, 1]$ and

$$d_l = d : x_l = x_l + \frac{\varepsilon}{2}(x_{l+1} - 2x_l + x_{l-1}) \tag{2.2}$$

with $0 \leq \varepsilon \leq 1$. In this case the mapping Φ can be written explicitly as follows

$$(\Phi\mathbf{x})_l = f(x_l) + \frac{\varepsilon}{2}(f(x_{l+1}) - 2f(x_l) + f(x_{l-1})) \tag{2.3}$$

Notice that this definition of CML, although corresponding to the historical case [15], cannot be identified as a discrete version of any continuum model (except in the trivial case where f_l is the identity mapping). In fact, any non-linearity or, equivalently, discontinuity of f_l prevents from the identification of a continuum limit.

Having this in mind we may introduce a seemingly more appropriate CML definition of the following type

$$(\Phi\mathbf{x})_l = f(x_l) + \frac{\varepsilon}{2}(x_{l+1} - 2x_l + x_{l-1}) \tag{2.4}$$

where the *reaction* and *diffusion* (coupling) terms do not act simultaneously on the state variable. Naively, one could interpret (2.4) as the space-time discrete version of the continuous reaction diffusion equation

$$\frac{dx}{dt} = \hat{f}(x) + \Delta x$$

with $\hat{f}(x) = f(x) - x$. The inconvenient of definition (2.4) is that $(\Phi\mathbf{x})_l$ may not necessarily belong to the original state space X_l for any given f_l (examples can be easily constructed). This implies the introduction of some constraints under the form of *modulo* operations to make (2.4) fit with the general definition of CML. Again, this makes doubtful any precise identification of the corresponding continuum equations.

In these lectures I shall refer mainly to the cases (2.3).

Anyway, for the sake of completeness, let me mention two further interesting examples.

The case of global coupling interaction can be interpreted as an evolution rule, where the local state variable is influenced by a sort of *mean field* term

$$(\Phi\mathbf{x})_l = f(x_l) + \varepsilon\left(-f(x_l) + \frac{1}{N-1}\sum_{m\neq l} f(x_m)\right) \tag{2.5}$$

where N is the number of nodes in Λ (which in general can also be infinite). Models like (2.5) have been introduced as *metaphores* of systems with *high connectivity*, e.g. biological networks, ecosystems and neural dynamics.

The final example is the so-called *open flow* CML

$$(\Phi\mathbf{x})_l = f(x_l) + \varepsilon\left(f(x_{l+1}) - f(x_l)\right) \tag{2.6}$$

which have been introduced as models for convective instabilities.

The considerations concerning the doubtful identification of a continuum limit exposed for the first two examples apply also to these last two cases.

What is also common to all these examples is that any *homogeneous state* $x_l = \bar{x} \in X_l, \forall l \in \Lambda$ is a solution of the dynamics independent of ε. Actually, the effect of the interaction term vanishes and the whole CML evolves like the isolated local map f_l. This implies that for any attractor of f_l there exists a corresponding attractor of the CML, with trivial, i.e. constant in space, structure. Notice that we did not say anything about the way such a *homogeneous state* may be reached when starting from a generic initial condition. In general we know that the state space A has a very intricate structure strongly dependent on N and ε; moreover, for fixed N, the relative stability of the CML attractors also depends on ε. Intuitively, one may expect that *homogeneous state* attractors may become stable for sufficiently high values of ε (provided the nonlinearity of f_l is not too strong). This is confirmed by numerical simulations.

3. Generic Behaviours in Specific Examples

In this Section I discuss in some detail two specific models of the type (1.3). Let me stress from the very beginning that most of the behaviours that I am going to describe are common to a wide class of models so that the restriction to these special cases does not represent a complete prejudice of generality to our speculations and conclusions.

A quantitative analysis of some of these generic properties will be the subject of the following Sections.

Here, my further purpose is to make the reader aware of the importance of having at disposal a clear interpretation of the phenomenological scenario as a pre-requisite for any conceptual progress.

3.1. COUPLED PERIODIC MAPS

A very simple version of a periodic map of the interval [0,1] into itself is the piece-wise linear transformation [7]

$$f(x) = \begin{cases} \lambda x \quad , \quad 0 \leq x \leq \lambda^{-1} \\ a + c(x - \lambda^{-1}) \quad , \quad \lambda^{-1} < y \leq 1 \end{cases} \tag{3.1}$$

with $a < \lambda^{-1}$ and c sufficiently small to guarantee the existence of a stable periodic orbit.

Let me first consider the *superstable* case $c = 0$. Map (3.1) has the following properties:

i) for the trajectory $\{f^t(x)\}$, $t \in Z_+$ of each $x \in (0,1]$ it exists $\tau(x) \geq 0$ such that $f^{\tau(x)}(x) = a$, $f^t(x) \neq a$ for $0 < t < \tau(x)$.
ii) for each pair of parameters (a, λ) there exists a unique number $k = k(a, \lambda)$ such that $f^k(a) = a$, $f^t(a) \neq a$ for $0 < t < k$.

This means that for any choice af a and λ we can selecte a superstable periodic solution of f with period k.

The results of numerical simulations for the lowest periodic cases ($k = 2,3$) of the corresponding CML defined in (2.3) show that even *trivial* local maps can exhibit rather complicated behaviours when diffusively coupled. In fact, starting from some randomly seeded initial condition (RSIC), after a few time steps various spatiotemporal patterns (clusters, states, interfaces) may appear. The dynamics is characterized by the presence of shock waves with different types of interfaces, kinks, intermittent fixed in space structures etc. These structures may also interact among themselves during a transient time needed to eventually reach a periodic attractor. The main feature of this model is that there is only a finite number of these patterns. This is due to the existence of globally attracting superstable periodic solutions of the local map f. Moreover, the number of patterns increases with the period of the superstable solution of f. Even the *complex* evolution ruled by the interaction of shock waves during the transient regime results in a finite number of patterns.

As well as the transient dynamics also the nature of the final attractor depends on ε. For instance, the *homogeneous state* attractor is typically selected for small values of ε, while for intermediate values of ε the attractors (despite the superstability of (3.1)) in general support a nontrivial space structure. It is worth mentioning that in this case shock waves during the transient separate different regions in space corresponding to some space-time periodic solution of the CML ; the corresponding final attractors may be space-time periodic or support travelling or static localized in space structures of the same kind of those observed during the transient (in the case of travelling shock waves time periodicity is only apparent due to the finite size of the chains used in numerical simulations). The periodic attractors can be classified into (m, n)-classes (from here on m and n will indicate periodicity in space and time, respectively). Formally, we include the periodic attractors with complex (non periodic) spatial structure in the special (∞, n)-class. (In numerical simulations (∞, n) solutions are never observed; it can be easily shown that the superstability of (3.1) does not allow for the existence of spatially irregular attractors).

Numerical simulations show two important facts:

i) the basin of attraction of any (m, n)-class solution shrinks for increasing values of both m and n;
ii) the transient to reach the asymptotic periodic attractor starting from a RSIC increases at most like a power of the number of nodes N in Λ.

This phenomenology takes place in some region of the parameter space for all CML with different local maps. In particular one must stress the abundance of solutions of (2,1)-class observed, for sufficiently large values of ε, in any CML of the type (2.3).

The presence of many space-time patterns is at the origin of the complex dynamical structure of CML, which is recovered also in the example under consideration, despite the extreme simplicity of the local map f_l. In view of this consideration, it is evident that a significant progress in understanding the complex pattern dynamics of CML deserves an appropriate technique to study the existence and the stability of the various space-time periodic solutions as a function of the parameters of the model (these points will be discussed in some detail in Section 5 for the case of superstable PCML; here I want to mention that a general rigorous approach to the stability analysis of the simplest classes of solution have been proposed in ref.[1]). Let me stress that CML are, by definition, dynamical systems with an infinite number of variables; as a consequence, the problem of the stability of a generic solution is non-trivial even for the simplest one, i.e. the (spatially) *homogeneous state*.

A richer scenario appears when one considers the local map f defined in (3.1) with $c > 0$, and sufficiently small to select a stable period-k solution. It is easy to show that the stability condition for such a solution fixes the upper bound $c < \lambda^{-k}$.

As it has been observed in numerical simulations (again with $k = 2, 3$) the corresponding CML dynamics acquires some new features with respect to the superstable case. These features illustrate other typical phenomena present in most CML with diffusive coupling interaction.

The most relevant among these features is that there exist finite regions in the parameter space where the transient toward a periodic attractor grows exponentially with N. From here on I shall indicate this dynamical regime as *chaotic transient* (CT). Let me notice that for the class of CML with local map given by (3.1) there is no possibility of chaotic (i.e. Lyapunov unstable) dynamics. Actually, diffusion cannot introduce any hyperbolicity mechanism when coupling stable and supertable periodic maps. Nonetheless, the space and time normalized auto-correlation functions,

$$\begin{aligned} C(i) &= | < x_{l+i}^t \quad x_l^t > / < x_l^t >^2 - 1| \\ C(\tau) &= | < x_i^{t+\tau} \quad x_i^t > / < x_i^t >^2 - 1| \end{aligned} \tag{3.2}$$

(where $< \cdot >$ denotes the spatiotemporal average and the dependence of the state variable on the discrete time index t is now explicit) decrease exponentially during CT dynamics. This is a typical feature of deterministic chaotic evolution (rapid loss of memory of the initial conditions).

In the other regions of the parameter space the pattern selection mechanism is similar to the one described in the superstable case, although, now, an infinite number of patterns has to be expected. The finite size of Λ together with the relative stability properties of the possible periodic attractors make extremely unprobable to observe the whole richness of this selection mechanism in numerical simulations.

This kind of mechanism is active also in the CT region. In this case spatially irregular patterns are favoured w.r.t. the periodic ones. Actually, the interfacial structures, that in the previous regime separate different spatially regular patterns remaining localized over a few number of nodes, tend to delocalize, giving rise to the just mentioned spatially irregular patterns, interlaced with localized regular ones. This mechanism accounts for the exponential decay of the space auto-correlation function. For what concerns the time auto-correlation function its exponential decay during CT is ruled by the intermittent mechanism of the domain boundaries separating different irregular patterns.

Let me also observe that we can infer from all of these considerations that (∞, n) solutions may exist on the infinite lattice Λ for parameter values in the CT region.

This scenario is strongly reminiscent of phase transitions; here, the different phases are selected by the competition mechanism between regular and irregular patterns. These qualitative remarks seem to suggest the possibility of characterizing critical behaviors in CML models in analogy with percolation phenomena. More precisely, spatially disordered patterns could be thought as an *active phase* competing with the *absorbing states* represented by the set of final periodic attractors. If such a correspondence could be formally established it could be reasonable to conjecture that phase transitions in CML are in the same universality classes of *contact processes* (including directed and dynamical percolation for multi-absorbing states). Recent numerical results on 1-d and 2-d models [9] indicate that this conjecture is correct at least in some restricted region of the parameter space. On the other hand there is also strong evidence that this is not a generic scenario for CML. Actually, in an extended region of parameter space the competition mechanism between *active* and *absorbing* phases is mediated by *nucleation* phenomena associated with a finite probability that some patch of the *absorbing* phases may spontaneously form inside the *active* region. Such nuclei may grow up to some finite average size (due to competition among nuclei of different *absorbing* states), or eventually be reabsorbed in the *active* phase. The occurrence of such finite size *ordered* structures inside the active phase prevents a clear identification of the critical behaviour and the phase transition appears as a *crossover* between two regions where the nucleation mechanism may or not dominate.

In the CT region it has been also observed sensitivity w.r.t. initial conditions in selecting the final periodic attractor, i.e. even a localized perturbation of the initial conditions may result in a drastic change of the CML asymptotic dynamics. This phenomenon claims for a definition of "chaos" on the infinite lattice in terms of the *unpredictability* of the final periodic attractor, due to the intrinsically nonlinear mechanism of localized front propagation. According to this observation "chaos" could be evaluated by measuring the *topological entropy* associated with these fronts, rather than on Lyapunov linear stability analysis. In fact, in CT one is allowed to define a quasi-stationary measure. Actually, numerical simulations show that during CT the maximum Lyapunov exponent is a well defined (i. e. well "relaxed") negative quantity, whose value depends on ε. As a consequence, Lyapunov stability analysis does not permit the identification of any sensitivity w.r.t. initial conditions in the usual sense.

One can guess that this peculiar representation of "chaos" in CML, i.e. *unpredictability* driven by non-linear dynamical mechanisms, has to be ascribed to the intricate structure in A of the basin boundaries of the many periodic attractors.

3.2. COUPLED CHAOTIC MAPS

The most intensively studied case is model (2.3) with

$$f(x) = ax(1-x) \tag{3.3}$$

which is the logistic map, with $a \in [0,4]$. Here I shall discuss the case $a > a_c = 3.569945$, where the map (3.3) is known to be chaotic, except for periodic windows of more or less small width. (Actually, a_c is the critical value of the nonlinear coupling a at which the sequence of period doubling bifurcations, originated by the stable fixed point of the logistic map, accumulate). For $a = 4$, moreover, the map (3.3) is proved to be mixing.

First of all let me observe that for $a < a_c$ the features of this CML are qualitatively equivalent to the ones discussed in the previous examples. This is not surprising, since in this case we are in presence of another example of a CML with diffusively coupled periodic maps.

For $a > a_c$ the CML model is characterized by a rich scenario when ε is varied in the interval [0,1]. For $\varepsilon = 0$ all periodic orbits are unstable and any RSIC gives rise to an asymptotic chaotic evolution. In the limit $\varepsilon \to 0^+$ it is known that the CML exhibit space-time chaos [5].

When ε is increased from 0 some space-time periodic states become stable. For instance, in various numerical simulations one frequently observes solutions of (2,2)-class. In other words, a transition from space-time chaotic to ordered motion seems to appear as ε varies. More recently, accurate numerical studies, based on a refined mathematical technique borrowed from low dimensional dynamical systems, have shown that between the space-time chaotic region and the range of stability of the (2,2)-solutions there are many other transitions corresponding to the stabilization of solutions of (m,n)-class with $m \geq 2$ and $n \geq 2$ [11].

I want to stress that the main difficulties encountered in numerical analysis of solutions with very high periodicity in space and time are mainly due to the following reasons:

i) the finite lattice size can be even smaller than the quite long spatial period of stable solutions;
ii) the very long transients needed to reach the asymptotic attractor.

In this sense, numerical simulations seem to indicate (uncorrectly) the persistence of a true chaotic regime even for values of ε where solutions of (m,n)-class should be stable.

In full analogy with the case of CML with stable periodic maps there are regions in the parameter space where CT are present. The same kind of sensitivity w.r.t. initial conditions introduced in the previous example applies to this model. The main difference in this case is that not only periodic, but also quasi-periodic and chaotic attractors may coexist even on finite size lattices.

Also in this case the intricate structure in A of the basin boundaries separating the final attractors seems to be the main property determining the *unpredictable* behaviour observed in the CT regime of CML. On the other hand, it is worth to stress that some differences are expected w.r.t. PCML. In particular, the existence of a submanifold in A (corresponding to the fully homogeneous state) where the motion is chaotic, may induce a fractal structure of basin boundaries, which is not expected in periodic CML.

4. Quantitative Analysis

In this section I aim to describe how to apply to CML some general tools that are commonly used to quantify the dynamical properties of SEDS.

4.1. LYAPUNOV SPECTRA

Lyapunov stability analysis is probably the most popular method used to characterize the phase-space structure of SEDS [3]. The method amounts to compute, in the infinite time limit, a spectrum of exponents resulting from the eigenvalue problem of the $N \times N$ matrix which rules the evolution in the tangent space of the dynamics. By definition N is the number of degrees of freedom, so that there are N Lyapunov characteristic exponents in the spectrum.

With reference to the CML version (2.3) the linearized evolution operator L_Φ in the tangent space reads

$$(L_\Phi u)_l = f'_l u_l + \frac{\varepsilon}{2}(f'_{l+1}u_{l+1} - 2f'_l u_l + f'_{l-1}u_{l-1}) \tag{4.1}$$

where u_l are the components of an N dimensional vector $\mathbf{u}$ defined on the tangent space of the state space A, and $f'_m = \frac{\partial f(x_m)}{\partial x_m}$ is computed along a trajectory of the dynamics (2.3).

Let me consider at time $T = 0$ an orthonormal basis of vectors $\mathbf{u}_1^0, \mathbf{u}_2^0, \cdots, \mathbf{u}_N^0$ in the tangent space and let me define the quantities

$$\eta_p = \lim_{T\to\infty} \frac{1}{T} ln \frac{\|\mathbf{u}_1^T \wedge \mathbf{u}_2^T \wedge \cdots \wedge \mathbf{u}_p^T\|}{\|\mathbf{u}_1^0 \wedge \mathbf{u}_2^0 \wedge \cdots \wedge \mathbf{u}_p^0\|} \quad , \quad p = 1, 2, \cdots, N \tag{4.2}$$

where $\wedge$ is the external product and $\|\cdot\|$ is the euclidean norm.

The p-th Lyapunov characteristic exponent is then given by

$$\lambda_p = \eta_p - \eta_{p-1} \tag{4.3}$$

The maximum Lyapunov characteristic exponent $\lambda_{\mathbf{max}} = \lambda_1$ corresponds to the case $p = 1$ (provided one defines $\eta_0 = 0$). A general criterion to establish the chaoticity of any SEDS is the positivity of $\lambda_{\mathbf{max}}$.

In view of the considerations reported in the previous Section, the Lyapunov spectra of CML may result in very different shapes when ε or the initial condition is varied. Moreover, in the CT numerical finite time estimates of $\lambda_{\mathbf{max}}$ seem to indicate that a quasistationary invariant measure should exist.

An alternative approach to the stability analysis of CML dynamics is the measure of the so-called *comoving* Lyapunov exponents [10]. These quantities are defined having in mind the spreading process of a local perturbation through the lattice. Although this definition does not rest on rigorous basis like (4.3), it is practically useful and strongly supported by numerical analysis. Operatively, one considers the evolution according to Φ of a point in A and at a certain arbitrary time, which can be fixed as the new origin on the time axis. At this moment of time the variables in the finite interval of node coordinates $I = [-L/2, L/2]$, centered around the origin are perturbed by an amount δ_i, $i \in I$. In the case (2.3) the perturbation can spread at most inside the *light cone* $-L/2-t < i < L/2+t$. It has been observed numerically (keeping track of both the perturbed and unperturbed evolution) that for sufficiently long times

$$|\delta_i^t| \simeq e^{\lambda_{\max}^c(v)t} \tag{4.4}$$

where $\lambda_{\max}^c$ is the maximum comoving Lyapunov exponent, which is a function of the velocity $v = i/t$. This means that this quantity is a measure of the exponential growth rate of a perturbation in a reference frame moving with velocity v. It is worth to observe that any estimate of the comoving exponents in the $t \to \infty$ limit would require to be measured on an infinite lattice, so that the finite width L of the initial perturbation can be considered an irrelevant parameter. Observe that this is no more true if, in analogy with the usual Lyapunov exponents definition, one would like to measure the whole spectrum of comoving exponents, since L determines the dimension of the subspace of the tangent space under investigation.

As a final remark I want to notice that linear stability analysis techniques in some cases may yield only partial informations about CML dynamics; for instance, there are examples where the maximum Lyapunov exponent is found to be strictly positive, while the evolution of a global quantity undergoes a quasi-periodic behaviour [8].

4.2. SYMBOLIC ENCODING

The phase transition-like behaviour discussed in the previous Section suggests intuitively that a statistical mechanics approach should be useful in characterizing some general features of CML.

In the absence of a precise knowledge about the possible existence of generating partitions of CML, one can approach the problem in a heuristic way . For instance, in the case of map (3.1) one could choose two symbols, say 0 and 1, to identify the state variable x_l when it is smaller or larger than λ^{-1}, respectively. The same symbolic reduction can be used for the map (3.3) when x_l is smaller or larger than 1/2.

In these cases the state variable x_l is substituted by a boolean variable b_l. (Notice that this may not be the most appropriate choice, when the CML dynamics, as it happens in some cases, exhibits a natural encoding with a larger number of symbols).

Various measure can be performed by this symbolic representation of CML dynamics. The analysis of the spatiotemporal patterns of model (2.3) equipped with the map (3.1) in the CT regime shows that this CML dynamics can be interpreted as an infinite memory Cellular Automaton (CA) rule. Actually, using the boolean encoding one can conclude that the ratio between undecidable and decidable space time configurations in the backward light-cone decreases exponentially. This analogy with CA is of great conceptual interest, because it singles out the main feature that differentiates space-time chaos w.r.t. usual low-dimensional deterministic chaos. In fact, CML in the CT regime, independently on the type of the final attractor, behaves like a CA complex rule (for instance rule 22 in the classification introduced by Wolfram [18]). For these models undecidability is originated by a fully nonlinear mechanism: changing only one bit amounts to propagate through the lattice, as time is increased, a disturbance which changes completely the dynamics w.r.t. the unperturbed case.

In CML, as we have already noted, the situation is quite similar: the spreading of localized perturbations is supported by wave fronts of various shapes, whose origin is related to the nonlinear properties of the local map.

Symbolic encoding can also be exploited for measurements of entropies of spatiotemporal patterns. More precisely, starting from an arbitrary node on the lattice, one can encode the spatial symbol sequence at time t as a real number $r_t = \sum_l b_l 2^{-l}$. It is evident that by this procedure the boolean representation of the space-time configuration is reduced to a time series $\{r_t\}$, which can be analysed by the usual embedding technique. In particular, one can construct an embedding space E_d of dimension d by evaluating r_t at subsequent times and defining a vector $\rho \in E_d$ as

$$\rho_d(t) = (r_t, r_{t+1}, \cdots, r_{t+d}) \qquad (4.5)$$

The fractal dimension associated to $\{r_t\}$ is then evaluated increasing the embedding dimension d. This is tantamount to computing the entropy (expressed in bits) of strings with increasing temporal heights. It has been shown [12] that an indication for spatiotemporal chaos is a linear increase of the entropy with the embedding dimension.

In order to perform a quantitative analysis a function which can be computed by the Grassberger-Procaccia algorithm [13] is the generalized information entropy

$$I_2(n) = \log_2 \frac{1}{N(N-1)} \sum_{i \neq j} \Theta(|\rho_i - \rho_j| - 2^{-n}) \tag{4.6}$$

where N is the total number of sampled points in the time series and Θ is the Heaviside function.

In the case of CML (2.3) equipped with local map (3.1) it has been found [7] that even in the CT regime I_2 tends to a limit value for finite values of d ($d \sim 4$). This shows that, despite the high complexity of the dynamics, this model does not exhibit (as expected) the peculiar features of space-time chaos, as it happens for the case (3.3) with a=4 and for small values of ε.

As a more general observation, let me notice that the boolean encoding of CML space-time evolution can be interpreted as a mapping of a dynamical configuration over an Ising-like spin configuration of statistical mechanics. (This analogy goes much further than expected: recently it has been found numerical evidence that some CML models exhibit true Ising phase transition behaviour in 2 (1+1) [4] and 3 (2+1) [16] dimensions.)

The crucial point of contact between these two descriptions is the following. Let me take into account a CML defined on an infinite lattice; beyond the mapping Φ generating the CML dynamics, there is also a dynamical system generated by the space translation operator S acting on the configuration space A, $S : A \to A$. In general one has $(Sx)_l = x_{l'}$, with $l, l' \in \Lambda$. For the simplest case ($\Lambda = Z$, i.e. infinite lattice in one space dimension) $(Sx)_l = x_{l+1}$, so that S represents the *left-shift* operator in space. The following step is to consider the space-time dynamics of the CML generated by the group of transformations (Φ, S) (which reduces to a semigroup if Φ is not time reversible) in the general framework of ergodic theory in order to define and characterize the various properties of CML by the language of statistical mechanics. For instance, in ref.[6] this approach was first introduced in order to define the conditions for space-time chaos in CML as the existence of a unique invariant measure μ in A, such that the dynamical system $(A, \mu, (\Phi, S))$ is the mixing one. On this basis, it has been also suggested that weak spatial interaction among chaotic maps should yield space-time chaos for the class of CML of the form

$$(\Phi x)_l = f(x_l) + \varepsilon\alpha(x_l)g_l(Fx_l) \quad (4.7)$$

where f is an expanding map of the interval [0,1] with a finite Markov partition, $\alpha(x_l)$ is a smooth function on [0,1], with $\alpha(0) = \alpha(1) = 0$ and the dependence of g_l on x_m decays exponentially as $|l - m| \to \infty$.

For the sake of brevity, I shall not report in this lectures the rigorous proof; I prefer to mention that this result has significantly contributed in clarifying the concept of space-time chaos; moreover, notwithstanding the special example to which it applies, it seems to extend its validity to many other CML of coupled chaotic maps.

4.3. SCALING ANALYSIS OF SELF-SIMILAR STRUCTURES IN STATE SPACE

In Section 2 we have already discussed that the intricate structure formed in A by the basins of attraction can be interpreted as another manifestation of space-time complex CML dynamics. A method to investigate quantitatively this fact is based on looking for some scaling laws apt to characterize the geometrical properties of A.

For this purpose, one can fix a RSIC for the state variables x_l and compute numerically the final attractor. (This kind of measurements can be performed for CML with relatively small size, due to computer time and memory limitations.) Then one can perturb the initial condition of a certain amount δ, $x_l \to x_l + \delta$, and compute the probability $P(\delta)$ that this new initial condition leads to a different attractor (in general, one can average over various RSIC). If one obtains the scaling law

$$P(\delta) \sim \delta^{\alpha} \quad (4.8)$$

with $0 < \alpha < 1$ one can conclude that the CML is characterized by fractal basin boundaries in A [14].

In spatio-temporal complex dynamics there are cases where the *uniformity exponent* $\alpha \sim 0$, meaning that even reducing the amplitude of the perturbation may result in almost no practical improvement in reducing the probability of finding the same attractor of the unperturbed RSIC. This is a sort of extreme situation, where A exhibits *riddled* basins [2]. This scenario can be characterized by extracting a new scaling law associated to a new exponent k [17].

Operatively, this kind of measurement can be performed by fixing a RSIC for all of the state variables x_l, except for x_0 (the state variable at the origin can be chosen without loss of generality if the CML has periodic boundary conditions). The state space X_0 is subdivided into disjoint intervals I_l of length a. One then assigns to all of these intervals the measure $\mu(I_l)$ and computes the quantity

$$\Omega(a) = \sum_l |\mu(I_l)| \tag{4.9}$$

which is supposed to behave as

$$\Omega(a) \sim a^{-k} \tag{4.10}$$

It remains to specify what is $\mu(I_l)$. One possible choice, also when various kind of attractors (chaotic, quasiperiodic and periodic) coexist, is to define

$$\mu(I_l) = \lambda_{\mathbf{max}}(I_l) - \bar{\lambda}_{\mathbf{max}} \tag{4.11}$$

The quantity $\bar{\lambda}_{\mathbf{max}}$ is the maximum Lyapunov exponent, averaged over initial conditions obtained by a very fine partition of X_0 which, in practice, defines $a_{\min}$; $\lambda_{\mathbf{max}}(I_l)$ is the maximum Lyapunov exponent of the initial condition taken at the interval I_l, corresponding to a partition of X_0 with $a > a_{\min}$.

5. Some Rigorous Results On Superstable CML

In the previous Sections I hope to have clarified which are the main points to be tackled in order to construct a satisfactory theoretical approach to CML. Unfortunately, only a few partial answers to the many questions that I have been raising can be stated on rigorous grounds. Here I want to report a series of results that have been inspired and, in some cases, supported by numerical analysis. Although they refer only to a specific model, I am convinced of their pedagogical interest.

Most of these results have been obtained in collaboration with L.A. Bunimovich, G.Martinez-Mekler and S. Ruffo (see ref. [7]).

5.1. STATES AND INTERFACES

Let me consider the CML (2.3) with local map (3.1) in the superstable case, $c = 0$.

For $\varepsilon = 0$ the local maps (3.1) are uncoupled. Therefore $\Phi = F$ and Φ^t maps for some $t \geq 0$ any nonzero coordinate x_i^0 into the interval $I_1 \equiv [\lambda^{-1}, 1]$. From that moment on the trajectory of the i-th coordinate is periodic, with a period $k \equiv k(a, \lambda)$.

First I want to determine for $\varepsilon \neq 0$ a condition such that for each coordinate $x_i^0 \in I_0 \equiv [0, \lambda^{-1})$, there exists a finite time $\tau_i(\varepsilon, a, \lambda; x_i^0)$ such that $x_i^{\tau_i} \in I_1$; in particular I will also estimate this time τ_i.

Let me define the transformations for $y \in (0, 1]$:

$$L(y) = (1 - \varepsilon) f(y)$$
$$U(y) = (1 - \varepsilon) f(y) + \varepsilon \tag{5.1}$$

It is easy to see that $U(y) > L(y)$. Moreover $(1 - \varepsilon) \geq L(y) \geq 0$ and $1 \geq U(y) \geq \varepsilon$, $\forall y \in [0, 1]$.

Let me also define the interval

$$\Delta_i(t) = [L^t(x_i^0), U^t(x_i^0)] \tag{5.2}$$

Lemma 1. For any $\varepsilon \in [0, 1]$, and $x_i^0 \in I_0$, if there exists $\bar{t} < \infty$ such that $U^{\bar{t}}(x_i^0) \in I_1$ and $U^t(x_i^0) \in I_0$ for $0 \leq t \leq \bar{t}$, then $F^t(x_i^0) \in \Delta_i(t)$ *for* $t < \bar{t}$.

Proof. According to (2.3) and (5.1) one has

$$U^t(x_i^0) = (1 - \varepsilon)\lambda((1 - \varepsilon)\lambda \cdots ((1 - \varepsilon)\lambda x_i^0 + \varepsilon) \cdots + \varepsilon) + \varepsilon \geq F^t(x_i^0) \tag{5.3}$$

and also

$$L^t(x_i^0) = ((1 - \varepsilon)\lambda)^t x_i^0 \leq F^t(x_i^0) \tag{5.4}$$

Both inequalities are satisfied since $U^t(x_i^0) \in I_0$ for all $t < \bar{t}$ and this is true also for $L^t(x_i^0)$. ■

Now one can prove the following

Lemma 2. If $(1-\varepsilon)\lambda > 1$, for any $\varepsilon \in [0,1]$ and $x_i^0 \in I_0$, there exists $\tau < \infty$ such that $x_i^\tau \in I_1$, $x_i^{\tau-1} \in I_0$.

Proof. $L^t(x_i^0)$ is an increasing function of t, provided $x_i^0 \in I_0$, $L(x_i^0) \in I_0, \dots, L^{t-1}(x_i^0) \in I_0$ and $(1-\varepsilon)\lambda > 1$. Therefore it exists $\bar{t} < \infty$ such that

$$L^{\bar{t}}(x_i^0) = ((1-\varepsilon)\lambda)^{\bar{t}} x_i^0 \geq \lambda^{-1} \tag{5.5}$$

This also implies the existence of $\tau \leq \bar{t}$ with the property $F^\tau(x_i^0) \in I_1$, $F^{\tau-1}(x_i^0) \in I_0$. ∎

Moreover one can obtain from (5.5) an upper bound for τ:

$$\tau \leq \bar{t} = [\log(\lambda^{-1}/x_i^0)/\log((1-\varepsilon)\lambda)] + 1 < \infty \tag{5.6}$$

where $[\cdot]$ denotes the integer part.

Analogously a lower bound for τ can be estimated by observing that as a consequence of Lemma 2 it exists $\hat{t}$ such that $U^{\hat{t}}(x_i^0) \geq \lambda^{-1}$. This implies the existence of $\tau \geq \hat{t}$ such that $F^\tau(x_i^0) \in I_1$. Making use of (5.3), one obtains

$$U^{\hat{t}}(x_i^0) = \alpha^{\hat{t}} x_i^0 + \varepsilon \sum_{n=0}^{\hat{t}-1} \alpha^n = \alpha^{\hat{t}} x_i^0 + \varepsilon \frac{1-\alpha^{\hat{t}}}{1-\alpha}$$

where $\alpha = (1-\varepsilon)\lambda > 1$. This allows to estimate $\hat{t}$ as

$$\hat{t} = [\log(\frac{\alpha(\alpha-1)+\varepsilon}{x_i^0(\alpha-1)+\varepsilon})/\log((\alpha)] < \infty \tag{5.7}$$

Lemma 2 implies that any coordinate x_i^0 enters I_1 after a finite time.
If these transient times are commensurable modulo k, the system relaxes asymptotically to a spatially homogeneous configuration. In the limit of infinite lattice the probability for this configuration to be the asymptotic state is zero.

Let us now consider the case $x_i^0 \in I_1$. Then

$$\begin{aligned} L^t(x_i^0) &= (1-\varepsilon)^t \lambda^{t-1} a \\ U^t(x_i^0) &= (1-\varepsilon)^t \lambda^{t-1} a + \varepsilon \frac{((1-\varepsilon)\lambda)^t - 1}{(1-\varepsilon)\lambda - 1} \end{aligned} \tag{5.8}$$

One can prove the following

Lemma 3. Given $x_i^0 \in I_1$, then both x_i^t and $\lambda^t a$ belong to Δ_i^t, provided $U^{t-1}(x_i^0) \in I_0$.

Proof. For what x_i^0 is concerned this result is a consequence of (5.8). The lower bound for $\lambda^t a$ follows from $\varepsilon \in [0,1]$. The upper bound can be verified immediately for $t = 1$ and $t = 2$. Let us suppose that it is true also for $t = l$, i.e. $\lambda^{-1} > U^l(x_i^0) \geq \lambda^l a$. By definition (5.1) one has

$$U^{l+1}(x_i^0) = \lambda U^l(x_i^0) + \varepsilon(1 - \lambda U^l(x_i^0)) > \lambda U^l(x_i^0) > \lambda^{l+1} a \qquad (5.9)$$

This proves the Lemma by induction. ■

Consider values of the parameters for which $\lambda^{k-1} a \in I_1$; this corresponds to the period k solution for the trajectory of each coordinate for $\varepsilon = 0$.

Lemma 4. If $\lambda^{k-1} a \in I_1$, the condition

$$L^{k-1}(x_i^0) > U^{k-2}(x_i^0) \quad , \quad x_i^0 \in I_1$$

guarantees that

$$L^l(x_i^0) > U^{l-1}(x_i^0) \quad , \quad (k-1) \geq l \geq 1$$

Proof. From the conditions of Lemma 4 we get easily $L^{k-1}(x_i^0) > U^{k-2}(x_i^0) - \varepsilon$. Writing down explicitly this formula and dividing both members of the inequality by $(1-\varepsilon)\lambda > 1$ one obtains

$$L^{k-2}(x_i^0) > U^{k-3}(x_i^0)$$

Applying this again the Lemma holds. ■

Corollary. If $U^{k-2}(x_i^0) \in I_0 \quad , \quad L^{k-1}(x_i^0) \in I_1$ then the $k-th$ image of $x_i^0 \in I_1$ enters I_1.

The proof follows immediately from Lemma 4.

The condition stated in Lemma 4 fixes an upper bound for ε such that $\Delta_i(t) \cap \Delta_i(t') = \emptyset$ for any $t \neq t'$.

Combining this with Lemma 3, one concludes that for $(1-\varepsilon)\lambda > 1$ the trajectory of any coordinate in the lattice will remain close to the period k orbit.

Let me also observe that the larger is k the smaller is ε in order to satisfy all the conditions discussed in this Section.

As an example let me consider the case $k = 2$. The condition of Lemma 4 is fulfilled if

$$0 \leq \varepsilon \leq \varepsilon_s = 1 + ((1-a) - ((1-a)^2 + 4\lambda a)^{\frac{1}{2}})/2\lambda a \tag{5.10}$$

Simple calculations show that this condition implies also $(1-\varepsilon)\lambda > 1$.

Moreover, since $U(x_i^0) \in I_0$ with $x_i^0 \in I_1$, it must be also true that

$$0 \leq \varepsilon \leq \varepsilon_r = \frac{(1-\lambda a)}{\lambda(1-a)}$$

If both conditions have to be valid ε must be smaller than $\min\{\varepsilon_s, \varepsilon_r\}$.

Let me consider the trajectory x_i^t of the coordinate x_i^0 at a moment $t = \bar{t}(x_i^0) + 2n, n \in \mathbf{Z}_+$, where n is chosen such that $t > \max\{\bar{t}(x_{i-1}^0), \bar{t}(x_{i+1}^0)\}$. If the conditions discussed in this example are fulfilled the trajectory of any coordinate is bounded from below by a and from above by $U_{\mathbf{max}} = (1-\varepsilon)^2\lambda a + \varepsilon$. It is also evident that asymptotically the trajectory of a point is inside the strip $S = [a, U_{\mathbf{max}}]^{\mathbf{Z}}$.

5.2. CLUSTERS OF HOMOGENEOUS SOLUTION AND THEIR INTERFACES

The CML dynamics has the following superstable solution for any $\varepsilon \in [0,1]$.

Definition 1. We call Homogeneous Solution(HS) the superstable solution of (2.3) where all coordinates x_i evolve synchronously in time according to the period k solution of the single map (3.1).

The spatially homogeneous configuration emerges if at some time t all coordinates enter I_1. Then the trajectory follows the periodic dynamics of map (3.1). The Jacobian of Φ evaluated for this solution has all zero eigenvalues.

Definition 2. We call Homogeneous Cluster(HC) any finite or semi-infinite sequence of coordinates taking the same value and evolving in time following the period k orbit of the single map (3.1). The k values of this periodic orbit are referred to as Phases of HC.

Different phases may coexist asymptotically on different parts of the lattice separated by coordinates belonging to **interfaces**. Restricting to period $k = 2$ we give here a definition of interface between HC's.

Definition 3. We shall say that a coordinate x_i belongs to the interface between HC's at time t if one of the following conditions hold

i) $f(x_i^{t-1}) \neq a, f(x_i^t) \neq a$;
ii) $f(x_i^{t-1}) = a, f(x_i^t) \neq \lambda a$ *or* $f(x_i^t) = a, f(x_i^{t+1}) \neq \lambda a$.

All the coordinates that at time t do not belong to moving or nonmoving interfaces belong at this time to a HC.

In the following we will restrict to the $k = 2$ case, i.e. $\lambda^{-2} < a < \lambda^{-1}$. We will also provide a method for computing explicitly the corresponding interfaces. They can be nonmoving ($v = 0$) or moving with rational velocity ($v = p/q$) ≤ 1 due to the space-time discreteness of the model. For the sake of simplicity we will discuss in detail only interfaces with short periodicity (small q), although the method can be extended to analyze any possible interface.

Lemma 5. If $k = 2$ it does not exist a nonmoving one coordinate interface between HC's.

Proof. Let x_0^0 be the only coordinate forming the interface between two HC's at some time that we fix at zero. Let us make the hypothesis that at an arbitrary time $2t \geq 2$ the previous point reproduces itself. After $2t - 1$ time steps only the coordinates $x_i^{2t-1}, -2t \leq i \leq 2t$ may not belong to HC's. Three different cases must be studied:

i) $x_i^{2t} = \lambda a, i \neq 0$, then $x_i^{2t-1} = a, \forall i$ and therefore at time $2t$ one has an HS, contradicting the hypothesis.

ii) $x_i^{2t} = a, i \neq 0$, then $x_i^{2t-1} \in I_1$ for $-2t \leq i \leq 2t$. As a consequence at time $2t$ one obtains an HS, again contradicting the hypothesis.
iii) without loss of generality we suppose that $x_i^{2t} = \lambda a, i > 0$ and $x_i^{2t} = a, i < 0$. The first requirement implies $x_0^{2t-1} = a$. Then, since $x_i^{2t-1} \in I_1, -2t \leq i \leq 1$, $x_{-1}^{2t} \neq a$, thus contradicting the hypothesis. ■

Observe that generalizing this reasoning it is also possible to prove that no single coordinate moving interface can exist.

Therefore, the smallest interface between HC's must contain at least two coordinates.

We follow a general approach to prove the existence of interfaces with prescribed properties and to obtain their analytical expression. We allow the presence of a certain number of "free" coordinates between HC's at some time t. We then follow the time evolution for a fixed number of time steps τ and we require that the configuration of time t is reproduced at time $t+\tau$ admitting that the interface may move in space. If $k=2$ then τ is a multiple of two. Let us define

$$\begin{aligned} \varepsilon_\alpha &= \frac{2(\lambda^2 a - 1)}{a\lambda(\lambda - 1)} \\ \varepsilon_\beta &= \frac{2(1 - \lambda a)}{a\lambda(\lambda - 1)} \end{aligned} \tag{5.11}$$

Lemma 6. For $k = 2$, $\varepsilon_\beta < \varepsilon < 1$ and $2/\lambda(\lambda+1) < a < 1/\lambda$ there exists a two coordinate, period two, $v = 0$ interface between HC's in different phases.

Proof. Let us consider the following initial point P^0: $x_i^0 = \lambda a$, $i < 0$, $x_i^0 = a, i \geq 2$ and some arbitrary values of the interval different from a and λa for x_0^0 and x_1^0. We require that the point reproduces itself at $t = 2$. The requirement imposes $x_0^2 = x_0^0$ and $x_1^2 = x_1^0$. One derives therefore $x_{-1}^1 = x_0^1 = a$ and x_1^1 and $x_2^1 \in I_1$. This allows the explicit computation of the two coordinates of the interface:

$$\begin{aligned} A = x_0^0 &= \lambda a - \frac{\varepsilon}{2} a(\lambda - 1) \\ B = x_1^0 &= a + \frac{\varepsilon}{2} a(\lambda - 1) \end{aligned} \tag{5.12}$$

In order to have $x_i^1 = a$ for $i \leq 0$ it is necessary that both x_0^0 and $x_1^0 \in I_1$; this leads to the constraint $\varepsilon_\beta < \varepsilon < 1$ which, in the allowed range of the parameters, holds only if $2/\lambda(\lambda+1) < a < 1/\lambda$. ■

Following the same approach it is straightforward to show that no $v = 0$ two coordinates interface can exist between HC's in the same phase. This results is easily generalized to higher velocities and larger periods.

In the study of interfaces of larger velocities two new critical values of the coupling constant appear.

$$\begin{aligned} \varepsilon_\gamma &= \frac{\lambda + 1 - \sqrt{(\lambda+1)^2 - 2\lambda\varepsilon_\beta}}{\lambda} \\ \varepsilon_\delta &= \frac{\sqrt{1 + 2\lambda\varepsilon_\beta} - 1}{\lambda} \end{aligned} \tag{5.13}$$

Lemma 7. For $k = 2$, $\max\{\varepsilon_\alpha, \varepsilon_\gamma\} < \varepsilon < 1$ and $\lambda^{-2} < a < 2/\lambda(\lambda+1)$ there exists a two coordinates period two, $v = 1$ interface between HC's in different phases.

Proof. Without loss of generality one can choose the same initial point P^0 as in the proof of Lemma 6. One then requires that the point at time $t = 2$ has the same structure of P^0 shifted by two space steps: e.g. $x_i^2 = \lambda a$ for $i < -2$, $x_i^2 = a$ for $i \geq 0$ and $x_{-1}^2 = x_1^0, x_{-2}^2 = x_0^0$. From this requirement one derives that $x_i^1 \in I_1$ for $-1 \leq i \leq 2$; this implies that $x_0^0 = A$ and $x_1^0 = B$ with A, B given in (5.12). Restricting to the case $A, B \in I_0$, which implies $1 > \varepsilon > \varepsilon_\alpha$ with $\lambda^{-2} < a < 2/\lambda(1+\lambda)$, one can verify that $x_0^1, x_1^1, x_2^1 \in I_1$ for any allowed values of ε, λ and a, while $x_{-1}^1 \in I_1$ only if $\varepsilon > \varepsilon_\gamma$. ■

A slight modification of the pattern leads to a $v = 1/2$ interface which can be shown to exist in two different regions of parameters: a)$1 > \varepsilon > \varepsilon_\alpha$ and $\lambda^{-2} < a < 2/\lambda(1+\lambda)$,b)$1 > \varepsilon > \varepsilon_\beta$ and $2/\lambda(1+\lambda) < a < \lambda^{-1}$.

Let us now consider the possible existence of two coordinate interfaces of periods higher than two.

Lemma 8. A $v = 0$ period four two coordinates interface between HC's cannot exist for any ε,λ and a.

Proof. Imposing again the condition that the initial point P^0 reproduces after four time steps one has $x_i^2 \in I_1$ for $-2 \le i \le 1$ and $x_i^3 \in I_1$ for $1 \le i \le 4$. Therefore, one can derive the values $x_0^0 = A$, $x_1^0 = B$. We have to consider three different situations

$$\begin{aligned} &(a) \quad A > B > \lambda^{-1} \\ &(b) \quad A > \lambda^{-1} > B \\ &(c) \quad \lambda^{-1} > A > B \end{aligned} \tag{5.14}$$

Case (a) implies $\varepsilon > \varepsilon_\beta$ and $\lambda^{-1} > a > 2/\lambda(1+\lambda)$. In this case one reduces to the period two $v = 0$ interface.

Case (b) implies $\varepsilon > \varepsilon_\alpha$ if $\lambda^{-2} < a < 2/\lambda(1+\lambda)$ or $\varepsilon < \varepsilon_\beta$ if $\lambda^{-1} > a > 2/\lambda(1+\lambda)$.

In the former case one obtains

$$x_2^1 = (1-\varepsilon)\lambda a + \frac{\varepsilon}{2}\lambda(a+B)$$

Since $B > a$ then $x_2^1 > \lambda a$ and $x_3^2 = a$. One also obtains

$$x_1^1 = \lambda a + \frac{\varepsilon}{2}a(\lambda-1)^2 - \frac{\varepsilon^2}{2}\lambda a(\lambda-1)$$

Let us suppose that $x_1^1 \in I_1$, then $x_2^2 = a$ and consequently $x_1^3 = B < \lambda^{-1}$, but x_1^3 should belong to I_1, therefore the hypothesis is contradicted. The other possibility $x_1^1 \in I_0$ implies $\varepsilon > \varepsilon_\alpha$ which again contradicts the hypothesis.

The other part of case (b) proves in a similar way contradicting the relation $\varepsilon > \varepsilon_\beta$.

Finally Case (c) implies $\varepsilon > \varepsilon_\alpha$ with $\lambda^{-2} < a < 2/\lambda(1+\lambda)$. As in the first part of Case (b) the condition $B \in I_0$ implies $x_2^1 > \lambda a$ and again one must have $x_1^1 \in I_0$. But computing explicitly x_1^1 one obtains $x_1^1 > \lambda a$, thus contradicting the previous condition. ■

The absence of two coordinates period four $v = 0$ interfaces does not exclude the existence of period four interfaces with a larger number of coordinates.

Lemma 9. If $\lambda^{-2} < a < 2/\lambda(1+\lambda)$ and $\varepsilon_\alpha < \varepsilon < \varepsilon_\gamma$ it exists a two coordinates period six interface between HC's in different phases.

Proof. Let us consider the initial point P^0, the conditions of the theorem imply $A, B \in I_0$. We exploit in this case the right-left space symmetry of the model and we require that the point symmetric to P^0 under space reflection is reproduced after three time steps. This leads to the pattern where $x^2_{-2}, x^2_{-1}, x^2_0, x^1_0, x^1_1$ and x^1_2 belong to I_1. The condition $x^2_{-2} \in I_1$ implies

$$\frac{\varepsilon}{2}(f(x^1_{-1}) - \lambda a) > \frac{1 - \lambda^2 a}{\lambda}$$

If $x^1_{-1} \in I_1$ this inequality implies $\varepsilon < \varepsilon_\alpha$ contradicting the hypothesis; then $x^1_{-1} \in I_0$. The condition $x^2_{-2} \in I_1$ leads to

$$\varepsilon^2(\lambda + 1 - \frac{\varepsilon}{2}\lambda) > -\frac{2\varepsilon_\alpha}{\lambda}$$

This inequality is always satisfied if $\lambda > -2$, which is true in our model. Therefore $x^2_{-2} \in I_1, \forall x^1_{-1} \in I_0$. This last condition reads

$$\varepsilon(\lambda + 1 - \frac{\varepsilon}{2}\lambda) < \varepsilon_\beta$$

and it is satisfied in the allowed region of parameters if $\varepsilon < \varepsilon_\gamma$. Once the conditions $\varepsilon_\alpha < \varepsilon < \varepsilon_\gamma$ are satisfied, all other coordinates in the interface satisfy the required bounds.

Due to the "left-right" reflection symmetry the point P^0 reproduces itself at time $t = 6$. ■

The explicit values in all two coordinates periodic interfaces between HC's in different phases can be calculated and agree with numerical simulations performed starting from random initial conditions. This implies that all these interfaces have a certain stability region.

Numerical simulations also show that interfaces of size larger than two exist and are stable. They can be computed analytically by the methods described in this Section but it is not an aim of this paper to give a detailed analysis of all possible interfaces. However a natural and interesting question is whether an upper bound on the size of the interface may exist.

Definition 4. We shall call block of the first (second) kind a sequence of k adjacent coordinates $x_i, x_{i+1}, \dots, x_{i+k-1}$ such that $a \leq x_j \leq \lambda^{-1}$ ($\lambda^{-1} < x_j \leq 1$) for $i \leq j \leq i+k-1$.

The following fact plays the central role in the dynamics of the CML under consideration

Lemma 10. Interfaces between homogeneous clusters cannot contain blocks of the first kind with lengths more than 6 and blocks of the second kind with lengths more than 4.

Proof. First we mention that, in view of Lemma 4, one can start the study of the evolution of an interface after the moment when all coordinates in a finite (but arbitrarily large) neighborhood of a given coordinate enter I_1. Just in this stage the corresponding time can be arbitrarily large. Then, as we shall see, the relaxation of an interface to a stationary state will take a finite time, bounded from above. We mention also that from the results of Subsection 5.1 it follows that, after entering I_1, the value of a coordinate will never be less than a.

Suppose now that at the moment $t-1$ there is a block of the second kind with $k=5$, then $f(x_j^{t-1}) = a$ for $i-2 \leq j \leq i+2$ and $x_j^t = a$ for $i-1 \leq j \leq i+1$. Therefore $f(x_i^{t+1}) = a$ and according to Definition 3 x_i^t does not belong to an interface. So we came to a contradiction and in an interface there cannot be more than 4 coordinates with values larger that λ^{-1}. Quite analogously one can check that in an interface there cannot be more than 6 coordinates with values between a and λ^{-1}. ■

Lemma 10 shows the main feature of CML generated by local maps with globally attracting superstable periodic trajectories and nearest-neighbour space interaction. In such systems there can be only a finite number of different patterns; in fact, perturbations of these patterns cannot propagate (in space and/or time) because of destruction by superstable periodic motion. Lemma 10 is obviously true not only for period 2, but also for one-dimensional maps (2.3) with superstable periodic trajectories of arbitrary period. Lengths of admissible blocks grow with the period of superstable motion.

5.3. ALTERNATE AND OTHER PERIODIC SOLUTIONS

Here we consider the solution that plays the central role in the dynamics of CML with nearest-neighbour diffusive coupling in the range of strong space interactions ($\varepsilon \sim 1$). This solution has the form of a standing wave with period 2 in space.

Denote

$$a_1 = \frac{\varepsilon a}{1-(1-\varepsilon)\lambda}$$
$$a_2 = \frac{a(1-\varepsilon-\lambda+2\varepsilon\lambda)}{1-(1-\varepsilon)\lambda} \tag{5.15}$$

Definition 5. We call Alternate Solution (AS), the configuration of the lattice where all even (odd) coordinates are equal to a_1 and all odd (even) coordinated are equal to a_2.

In analogy with Homogeneous Clusters one defines Alternate Clusters (AC's). Denote

$$\varepsilon_\eta = \frac{\lambda-1}{\lambda(1-a)} \tag{5.16}$$

Lemma 11. For $\varepsilon > \varepsilon_\eta$ the Alternate Solution exists and is linearly stable.

Proof. We impose the fixed point solution for the model (2.3), requiring that odd(even) coordinates take the same constant value. We obtain consequently that $x_{2i} = a_1, x_{2i+1} = a_2$. This solution exists only if $1 > a_2 > \lambda_{-1} > a_1 > 0$. Among these bounds the most strict one is $a_1 < \lambda^{-1}$ which gives $\varepsilon > \varepsilon_\eta$. One can also study the stability of the AS; the Jacobian of the map (3.1) evaluated for N coordinates on the AS has $N/2$ zero eigenvalues and $N/2$ equal to $(1-\varepsilon)\lambda$. Therefore the local stability condition implies $\varepsilon > (\lambda-1)/\lambda$. However the previous bound $\varepsilon > \varepsilon_\eta$ is more strict than this latter one. ■

Observe that the two values, a_1 and a_2, of the AS converge to the period 2 solution of the local map as $\varepsilon \to 1_-$. Moreover the first correction near $\varepsilon = 1$ for both values has the same coefficient $a(\lambda-1)$.

When HC's and AC's coexist, *interfaces* between them will form. Numerical simulations show that they can be of various shapes, size and velocity. We study here analytically only one of the simplest, which happens to be the most frequently observed in numerical simulations starting with random initial points. It is the $v = 1$ one coordinate interface between a HC and an AC.

Let

$$\varepsilon_\phi = \frac{2(\lambda - 1)(1 - \lambda a)}{\lambda(2 - 3\lambda a + a)} \tag{5.17}$$

and

$$a^* = \frac{2}{3\lambda - 1} \tag{5.18}$$

Lemma 12. For $1 > \varepsilon > \varepsilon_\phi > 0$ and $2/\lambda(1+\lambda) > a > \lambda^{-2}$ it exist a $v = 1$ one coordinate interface between a HC and an AC.

Proof. Consider the following initial point Q^0: $x_i^0 = a$ for $i < 0$,$x_{2i+1}^0 = a_1$ and $x_{2i+2}^0 = a_2$ for $i \geq 0$ and some arbitrary value different from $a, \lambda a, a1, a2$ for x_0^0. We require that Q^0 reproduces after two time steps shifted by two space steps to the right: e.g. $x_i^2 = a$ for $x < 2$,$x_{2i+1}^2 = a_1$ and $x_{2i+2}^2 = a_2$ for $i > 0$, while $x_2^2 = x_0^0$. This implies that $x_i^1 \in I_1$ for $-1 \leq i \leq 1$. Let us now suppose that $x_0^0 \in I_1$, then $x_1^1 = (1-\varepsilon)\lambda a_1 + \varepsilon a$. The condition $x_1^1 \in I_1$ is satisfied only if $\varepsilon < \varepsilon_\eta$ which is incompatible with the range of value where the AS exists. Therefore we must have $x_0^0 \in I_0$. Computing now explicitly x_0^0 and imposing this last condition one obtains

$$\varepsilon > \varepsilon_\phi \quad \mathit{for} \quad a < a^*$$

$$\varepsilon < \varepsilon_\phi \quad \mathit{for} \quad a > a^*$$

The intersection of these values of ε with the interval $[0, 1]$ is non empty only if the conditions of the Lemma are satisfied. Automatically the bound $x_{-1}^1, x_0^1 \in I_1$ is satisfied. ■

One can consider interfaces that separate AC's or AC's and HC's in the same manner as in Subsection 5.2 for interfaces between HC's. These interfaces are formed by coordinates x_i that do not belong simultaneously both to AC's an HC's. Therefore the following Lemma is an immediate consequence of Lemma 10.

Lemma 13. Interfaces between AC's or between an HC and a AC can not contain blocks of the first kind with length $k > 6$ and blocks of the second kind with $k > 4$.

It is easy to verify that the pattern symmetric to the one described in Lemma 12 under "left-right" reflection does not exist. However an interface propagating to the right inside a HC with $v = 1$ and coexisting with an AC is often observed in numerical simulations.

In full generality we consider the initial point P^0 with $x_i^0 = \lambda a, \quad 1 \leq i \leq \infty$ and $x_0^0 \in I_0$. In the following Lemma, whose proof is a simple algebraic exercise, we do not require any condition for x_i^0 with $i < 0$.

Lemma 14. For the initial point P^0 if $x_2^2 = x_0^0$ and $1 > \varepsilon > \max\{\varepsilon_\alpha, \varepsilon_\gamma\}$ then $x_1^1 = a + \frac{\varepsilon}{2}(\lambda x_0^0 - a) \in I_1$ with $x_0^0 = A$.

This Lemma establishes the conditions under which the rightmost coordinate of the interface propagates inside the HC.

The coexistence with the AC requires further conditions on P^0, i.e. $x_{2i}^0 \in I_0$ and $x_{2i+1}^0 \in I_1$ for $i < 0$.

Lemma 15. Given the initial point P^0 if $1 > \varepsilon > \max\{\varepsilon_\alpha, \varepsilon_\gamma\}$, $x_i^2 = x_{i-2}^0 \quad \forall i$ and $x_{2i}^1 \in I_0$, $x_{2i-1}^1 \in I_1$, $i \leq 0$ then $x^t_{-2i} \to a_1$ and $x^t_{-2i-1} \to a_2$ as $i \to \infty$, for $t \geq 0$.

Proof. Under the hypothesis of the Lemma one obtains

$$\begin{aligned} x^2_{-2i} &= (1-\varepsilon)\lambda x^1_{-2i} + \varepsilon a \quad i \geq 0 \\ x^1_{-2i} &= (1-\varepsilon)\lambda x^0_{-2i} + \varepsilon a \quad i \geq 0 \end{aligned} \tag{5.19}$$

These equations define the following space map

$$x^0_{-2i-2} = f o f(x^0_{-2i}) \tag{5.20}$$

with

$$f(y) = (1-\varepsilon)\lambda y + \varepsilon a \tag{5.21}$$

The map (5.20) has a stable fixed point in a_1 in the region of parameters where the AS exists and is linearly stable (see Lemma 11). This proves the first part of the Lemma. Moreover

$$x^2_{-2i+1} = (1-\varepsilon)a + \frac{\varepsilon}{2}\lambda(x^1_{-2i} + x^1_{-2i+2}) \quad i \geq 1 \tag{5.22}$$

Since the fixed point of Eq. (5.20) is reached, then asymptotically $x^0_{-2i-1} = x^2_{-2i+1} \to a_2$ for $i \to \infty$. We have restricted the time interval to $0 \leq t \leq 2$, but the reasoning extends for any time. ∎

The mechanism we have described allows to reach the Alternate Solution asymptotically by a space translation to the left of the interface with a HC. The Alternate Solution is reached exponentially fast. The two characteristic values of the Alternate Solution are reached from below or from above depending whether $x^0_0 < a_1$ or $\lambda^{-1} > x^0_0 > a_1$. The second condition corresponds to $\varepsilon \in (\varepsilon_-, \varepsilon_+)$, while the first one is satisfied outside this interval, where

$$\varepsilon_{\pm} = \frac{3\lambda + 1 \pm \sqrt{\lambda^2 + 6\lambda + 1}}{2\lambda} \tag{5.23}$$

5.5. MORE COMPLEX PATTERNS

The main Lemma 10 does not exclude the existence of long interfaces between AC's. These long interfaces were never observed in computer simulations. However, it is possible to show that they exist; one can construct them simply by glueing together consecutively blocks of the first and of the second kind (see Definition 4). These long interfaces are formed by clusters that correspond to solutions of CML with longer periods.

Definition 6. We shall say that a solution (state) $x = \{x_i\}, \quad -\infty < i < \infty$, of CML belong to (m,n)-class if it has (minimal) period m in space and (minimal) period n in time, i.e. $x_i = x_{i+m}, \quad -\infty < i < \infty$, and $(\Phi^n x)_i = x_i, \quad -\infty < i < \infty$,.

According to this Definition the Homogeneous Solution belongs to the class (1,1) and the Alternate Solution belongs to the class (2,1).

For any prescribed sequence of blocks of the first and of the second kind, alternating with different lengths, it is possible to check whether it exists the solution of CML with corresponding periodicity in space and with some fixed period in time. An answer to such a question require an elementary but tedious algebra.

As an example we consider here periodic solutions of class (3,2) only. Denote the values of coordinates of this solution at some fixed moment of time by a_1, a_2, a_3. In view of the results of Subsections 5.2 and 5.3 we can assume that $a_j \geq a, \quad j = 1, 2, 3$. It is easy to check that (3,2) solution exists if and only if $a_1 = a_2$ (or $a_2 = a_3$). Suppose that $a_2 = a_3$, then four situations are possible:

i) $a_1 > \lambda^{-1}, \quad a_2 > \lambda^{-1}$: this case leads to a Homogeneous Solution , i.e. a class (1,1) pattern;
ii) $a < a_1 < \lambda^{-1}, \quad a < a_2 < \lambda^{-1}$: the (3,1) solution corresponding to this case exists for $1 < \lambda < \sqrt{3}$;
iii) $a_1 > \lambda^{-1}, \quad a < a_2 < \lambda^{-1}$: the (3,1) solution exists if

$$\varepsilon < \inf\{\frac{2}{3}, \frac{\sqrt{2}}{\lambda}, \varepsilon_\beta, \varepsilon_\eta, \frac{\varepsilon_\alpha}{2}, \frac{\varepsilon_r}{2}, \hat{\varepsilon}\}$$

where

$$\hat{\varepsilon} = \frac{a(1-\lambda) + \sqrt{a^2(1-\lambda)^2 - 2(1-a)(\lambda a - 1)}}{\lambda(1-a)}$$

iv) $a < a_1 < \lambda^{-1}, \quad a_2 > \lambda^{-1}$: in this case the solution (3,1) does not exist.

In general clusters of (m_1, n_1) and of (m_2, n_2) solutions can be separated by interfaces that are formed by clusters of (r, s) solutions, where $r < \min\{m_1, m_2\}$.

The basins of attraction of (m, n) -solutions shrink with the increase of their space period m. In particular, no solutions with $m \geq 3$ were observed as asymptotic states in computer experiments dealing with random initial conditions.

References

[1] Afraimovich, V.S., L.A. Bunimovich, Random and Computational Dynamics **1**, 423 (1993).

[2] Alexander, J.C., J.A. Yorke, Z. You, I. Kan, *Int. J. Bif. & Chaos* **2**, 795 (1992).

[3] Benettin, G., L. Galgani, A. Giorgilli J.M. Strelcyn, *Meccanica* **9**, 21 (1980).

[4] Bunimovich, L.A., private communication.

[5] Bunimovich, L.A., Ya. G. Sinai, in *Theory and Applications of Coupled Map Lattices*, K. Kaneko editor, J. Wiley (1993).

[6] Bunimovich, L.A., Ya. G. Sinai, *Nonlinearity* **1**, 491 (1988).

[7] Bunimovich, L.A., R. Livi, G. Martinez-Mekler, S. Ruffo, *Chaos* **2**, 283 (1992); A. Politi, R. Livi, G.L. Oppo, R. Kapral, *Europhys. Lett.* **22**, 571 (1993); R. Kapral, R. Livi, G.L. Oppo, A. Politi, *Phys. Rev.* **E49**, 2009 (1994).

[8] Chaté, H., P. Manneville, *Prog. Theor. Phys.* **87**, 1 (1992); *Europhys. Lett.* **14**, 409 (1991); J.A.C. Gallas, P. Grassberger, H.J. Herrmann, P. Ueberholz, *Physica* **A180**, 19 (1992).

[9] Cuche, Y., R. Livi, A. Politi, in preparation.

[10] Deissler, R.J., K. Kaneko, *Phys. Lett.* **A119**, 397 (1987).

[11] Giberti, C., C. Vernia, *Chaos* **4** (1994), to appear.

[12] Grassberger, P., *Phys. Scr.* **40**, 346 (1989).

[13] Grassberger, P., I. Procaccia, *Phys. Rev. Lett.* **50**, 346 (1983).

[14] Grebogi, C., S.W. McDonald, J.A. Yorke, *Phys. Lett.* **A99**, 415 (1983); S.W. McDonald, C. Grebogi, E. Ott, J.A. Yorke, *Physica* **D17**, 125 (1985).

[15] Kaneko, K., *Progr. Theor. Phys.* **72**, 480 (1984); I. Waller, R. Kapral, *Phys. Rev.* **A30**, 2047 (1984); J. Crutchfield, *Physica* **D10**, 229 (1984); S.P. Kuznetsov, A.S. Pikovsky, *Izvestija VUZ, Radifizika* **28**, 308 (1985).

[16] Miller, J., D.A. Huse, *Phys. Rev.* **E48**, 2528 (1993).

[17] Ott, E., Y. Du, K.R. Sreenivasan, A. Juneja, A.K. Suri, *Phys Rev. Lett.* **69** (1993); Y. Du, E. Ott, *Physica* **D67**, 387 (1993).

[18] Wolfram, S., Theory and Applications of Cellular Automata, World Scientific, Singapore (1986).

SOME DYNAMICAL PROPERTIES OF ONE-DIMENSIONAL CELLULAR AUTOMATA

ALEJANDRO MAASS
Universidad de Chile
F.C.F.M.
Departamento de Ingeniería Matemática
Casilla 170-3 Correo 3, Santiago
Chile
amaass@dim.uchile.cl

ABSTRACT. In this note we review some results about the ergodic theory and topological dynamics of one-dimensional cellular automata. First, we describe those classifications of cellular automata with respect to their attractors and equicontinuous points. In section 4 we study onto cellular automata, in particular we give some results concerning to positively expansive cellular automata. Finally, we describe some symbolic dynamics of the limit sets of cellular automata.

1. Introduction

Let A be a finite alphabet and d be a positive integer. We denote by A^{K_d}, where K_d is $\mathbb{Z}^d$ or $\mathbb{N}^d$, the set of maps $x : K_d \to A$; in the classical literature concerning cellular automata we call them *configurations* and A^{K_d} is called a *full shift*. For each $i \in K_d$, $x \in A^{K_d}$, $x(i)$ (or simply x_i) denotes the value assigned to the coordinate i of x.

A d–dimensional cellular automaton is a discrete dynamical system $F : A^{K_d} \to A^{K_d}$ defined by:

$$\forall\ x \in A^{K_d}, i \in K_d \qquad F(x)_i = f(x_{i+v_1}, ..., x_{i+v_m}),$$

where the map $f : A^m \to A$ and vectors $v_1, ..., v_m \in K_d$ are given. We say that f is the *local rule* of the automaton and that $V(i) = (i + v_1, ..., i + v_m)$ is the neigbourhood of the coordinate $i \in K_d$.

From a purely descriptive point of view, the dynamics of a cellular automaton consists of update synchronously the values of each coordinate of a given configuration following the local rule f, which depends for each cell of a finite number of neighbours Therefore, the dynamics of cellular automata are homogeneous in the configuration space. This kind of dynamics are specially adapted to *natural* processes that can be described by a massive collection of simple objects which interact locally.

In this note we shall be specially interested in the dynamics of one-dimensional cellular automata. In this case, the neigbourhood of the cell $i \in K_1$ is: $V(i) = \{i + l, ..., i + r\} \subseteq K_1$ where $l, r \in K_1$ and $l \leq r$. We shall always assume that $V(0) = \{l, ..., r\}$ is minimal, that is, the local rule depends explicitly of the coordinates $l + i$ and $r + i$. It is necessary to point out that a cellular automaton can be defined by several local rules f where the parameters r and l are different. So, we can define the neigbourhood of a cellular automaton by only consider one parameter $r \in \mathbb{N}$, called the *radius* of F, such that for each $i \in K_1$, $V(i) = \{i - r, ..., i, ..., i + r\}$ if $K_1 = \mathbb{Z}$, and $V(i) = \{i, ..., i + r\}$ if $K_1 = \mathbb{N}$; of course, both intervals are not necessarily minimal.

The study of cellular automata was introduced by John Von Neumann at the end of the forties. The main motivation was to construct *simple* mathematical models, in the one hand capable of universal computations, and on the other hand capable of self-reproduction. Since then, the study of the dynamics of cellular automata has been followed from different fields: biology, computer science and physics. The purpose is to make discrete simulations of natural processes called *complex*. The development of high performance computer systems and also the apparition of computers with several processors working in parallel (Connection Machine, CRAY, and others) have contribute, with no doubt, to the popularity of cellular automata. Both local and parallel nature of cellular automata make very easy their implementation in a computer device. Nowadays, we can find computer systems capable of simulate these processes almost in *real time*.

Among the processes that have been studied using simulations with cellular automata we can find in physics: spin glasses systems, reaction–diffusion processes, various phenomena in turbulence theory; in biology, we have studied the tumours growth, the excitement of muscular tissue; in theoretical computer science we have tried fundamentaly to find cellular automata capable of *complex* computations. Thus, we can find cellular automata which can simulate the operations of classical logic, or simply a Turing machine, even universal. It is always amazing to see how cellular automata conceived for simulation of *natural* systems appearing in domains quite far–off the computer science are capable of universal computations. One of the most remarkable examples is the so called "game of life" defined by Conway in 1970.

In the classical theory of cellular automata we often say that those systems which simulate universal devices are very *complex*. However, we have to point out that in the literature the words "complex" and "complexity" have multiple meanings, sometimes contradictories; up to now there is not a unified notion of "complexity" serving to classify cellular automata.

In this article we shall present some recent results concerning those aspects of the *complex* behaviour of cellular automata that follow from their interpretation as topological flows: cellular automata are seen as dynamical systems given by a compact metric space $X = A^{K_1}$ endowed with a continuous transformation $T : X \to X$. Also, whenever T is onto we can consider an invariant measure μ (that always exists). From ergodic theory point of view, it is important to prove the ergodicity, or some mixing properties with respect to the invariant measure, or eventually to prove that they are conjugate with Bernoulli systems or others. From a purely topological point of view, we also consider topological notions of mixing, we try to compute some topological invariants such as entropy or the dimension group. It is also important to consider equicontinuity properties like sensibility on initial conditions or others. Finally, the symbolic dynamics of cellular automata consists of describe the subshifts that appear with the dynamics of the automata.

The study of cellular automata from a purely mathematical point of view begin in 1969 with the article of Hedlund [19]. He mainly consider questions about one dimensional cellular automata; he call them *endomorphisms of a full shift*. The work of Hedlund was motivated by problems appearing at that time in symbolic dynamics: combinatorial characterization of onto cellular automata, or of the automorphism group of a full shift, etc. Thus, his work has motivate several works in this last domain, among them: [3], [11], [37], [38], [39].

The definitive impulse to the study of cellular automata from a mathematical point of view has been done by the program stated by S. Wolfram at the beginning of the eighties [48] [49]. By means of computer simulations, he gave an "empirical" classification of cellular automata, following what appear in the computer screen of their asymptotic behaviour. From then, the study of cellular automata has been focused to some dynamical aspects: ergodic theory, topological dynamics, symbolic dynamics. Most of later works concerning cellular automata, however different they can be, are fundamentaly oriented to give a mathematical sense to Wolfram's classification. Among the main contributions to the study of the dynamics of cellular automata we can mention: [10], [12], [16], [17], [20], [21], [22], [23], [24], [25], [26], [31], [32], [33], [36], [46], [43], [44], [45].

At the beginning of the nineties Jarkko Kari published a serie of remarkable results concerning the undecidability of several properties of cellular automata. The best reference is his Phd Thesis [28]. Let us recall that a problem is decidable if there exist an algorithm which receives as input the codification of a cellular automaton and answer "YES" if it has a desired property and "NO" otherwise. The first result of Kari, which give the answer to a conjecture that was opened for more that 15 years, stated that for cellular automata of dimension greater than one the surjectivity and injectivity problems are undecidable (we point out that in the one dimensional case both problems are decidable). He also proved that any "non trivial" question concerning the asymptotic dynamics of cellular automata is undecidable [27]. After these strong results there are at least two possible directions to follow. First of all it would be nice to distinguish classes of cellular automata where the problems described below become decidable and moreover where we can describe completely their dynamics. On the other hand, cellular automata do not present all dynamical behaviour that we find in the nature (we will see that in the sequel).

2. Definitions and Notations

Most of the properties and concepts given in this section are standard in ergodic theory and topological dynamics. Motivated readers can also see [13].

Let us begin with some general facts of topological dynamics and ergodic theory. A topological *flow* is a compact metric space X endowed with a surjective endomorphism T. Consider two flows (X,T) and (X',T'). We say that (X',T') is a factor of (X,T) (or (X,T) is a cover of (X',T')) if there exists a continuous onto map $\pi : X \to X'$ such that $T' \circ \pi = \pi \circ T$; if π is also 1-to-1 (X,T) and (X',T') are said to be conjugate. Let μ be a T-invariant measure, that is $T\mu(A) = \mu(T^{-1}(A)) = \mu(A)$ for any Borel set A of X. We endow (X',T') with the invariant measure μ' defined by $\mu'(A') = \pi\mu(A') = \mu(\pi^{-1}(A'))$ for any Borel set A' of X'. In the category of measurable dynamical systems, where topological structures are replaced by measurable ones, the natural notion of equivalence between systems is the *isomorphism*, that is, the existence of a bimeasurable bijection defined over a set of measure one, exchanging the measures and transformations.

A topological flow (X,T) is transitive if for any pair of open sets U, V of X there is a positive integer n such that $U \cap T^{-n}V \neq \phi$. Moreover, if there is $n_0 \in \mathbb{N}$ (only depending on U and V) such that for all $n \geq n_0$, $U \cap T^{-n}V$ is not empty, we say that the system is topologically mixing or simply *mixing*. The mixing and transitivity properties are conjugacy invariants. Transitivity is intimately related to ergodicity. Recall that a flow (X,T,μ), where μ is T–invariant, is ergodic (with respect to μ) if the invariant sets of X are μ equals to X or $\emptyset$.

Proposition 2.1. Let (X,T) be a flow and suppose that μ is a T–invariant measure whose support is X. If (X,T) is ergodic then it is topologically transitive. ■

In this article X is always a symbolic space. Let A be a finite alphabet. We denote by A^* the set of finite sequences or *words* on A, including the *empty word* λ; in other words, $A^* = \cup_{n\in\mathbb{N}} A^n$ where A^n is the set of words of length n of A^*; $|w|$ denotes the length of the word $w \in A^*$. A *language* L is an arbitrary subset of A^*; we denote by L_n the set of words of length n of L.

Let us put $K = K_1$. A^K is the set of infinite sequences $x = (x_i)_{i\in K}$, where $x_i \in A$; we call them *configurations*. For $i \leq j$ in K put $x(i,j) = x_i...x_j$. Let x be a configuration and $w = w_0 \ldots w_{n-1}$ be a word of length n, we denote by wx the configuration y defined by $y(0, n-1) = w$ and $y_{n+i} = x_n$, $i \in \mathbb{N}$. If w is a word of length n, then w^m is the word of length mn formed by repeating w m times. We shall use w^∞ to denote the string of symbols formed by repeating infinitely many times w to the right of some prescribed position (analogously we define ${}^\infty w$). Let $x = (x_i)_{i<0}$, $y = (y_i)_{i\geq 0}$, and $l \in \mathbb{Z}$, the configuration $z = x|_l y = (z_i)_{i\in\mathbb{Z}}$ is defined by : $z_i = x_{i-l}$ if $i < l$ and $z_i = y_{i-l}$ if $i \geq l$.

A^K is endowed with the product topology and the *shift* $\sigma : A^K \to A^K$, $\sigma(x) = (x_{i+1})_{i\in K}$. When $K = \mathbb{Z}$, σ is an homeomorphism. The family of cylinder sets $[w]_i = \{x \in A^K / x(i, i+|w|-1) = w\}$, where $w \in A^*$ and $i \in K$, is a fundamental base of clopen (for close and open) neighbourhoods of A^K. For this topology A^K is a compact metric space; the distance is defined by $d(x,y) = \sum_{i\in K} \frac{d_i(x,y)}{2^{|i|+1}}$, where $d_i(x,y) = 1$ if $x_i \neq y_i$ and 0 otherwise; two configurations are close with respect to this distance if for some large n their coordinates coincide from $-n$ to $+n$ if $K = \mathbb{Z}$ or from 0 to $+n$ if $K = \mathbb{N}$. The flow (A^K, σ) is called one-sided *full shift* when $K = \mathbb{N}$, or simply *full shift* when $K = \mathbb{Z}$.

A *subshift* or *symbolic flow* is a closed shift-invariant subset S of A^K endowed with σ. In general we identify the flow (S, σ) with the space S. The language associated to the subshift S is $L(S) = \{w \in A^* / \exists\ x \in S,\ i \in K,$ such that $x(i, i+|w|-1) = w\}$. It is well known that S, whether oner- or two-sided, is completely described by its language, and a one-sided subshift is the projection of a two-sided one.

In this article *subshifts of finite type* (SFT) and *sofic systems* play a prominent role. A subshift $S \subseteq A^K$ is said to be of finite type if there exist a positive integer N (assumed to be minimal), and a collection L of words of length N, such that $x \in S$ if and only if $x(i, i+N-1) \in L$ for all $i \in K$. The integer N is called the *order* of S. Any SFT S of order N is conjugate to a SFT of order 2 (called Markov system), by the map which associates to each $x \in S$ the point $y = (x(i, i+N-2))_{i\in K} \in L_{N-1}(S)^K$. A Markov system, $S \subseteq A^K$, is associated to an incidence matrix $M(S)$ indexed by A such that the entry corresponding to $(a,b) \in A^2$ is equal to 1 if $ab \in L_2(S)$, 0 otherwise. The mixing properties of the flow (S, σ) are deduced from properties of its transition matrix. Thus, S is transitive if for all $(a, b) \in A^2$ there exists $n \in \mathbb{N}$ such that $M(S)^n(a, b) > 0$, and S is mixing if n can be chosen independently of the entry (a, b); n is called a transition length of S.

Let T be a subshift and N be a positive integer. The approximation of order N of T is the SFT of order N, T_N, such that $L_n(T_N) = L_n(T)$ for all $1 \leq n \leq N$. It is clear that T_N contains T and that $T = \bigcap_{N \geq 1} T_N$. When T is mixing, T_N is mixing.

A subshift S is a *sofic system* if it is the factor of a SFT [47]. Any SFT is a sofic system (as a factor of itself), but sofic systems are much more general. Sofic systems can be covered by different SFT's, in this paper we only consider the *minimal Shannon cover* for transitive sofic systems defined as follows (for the proofs see [14]):

Let S be a transitive sofic system. There exists a finite automaton $\mathcal{A}(S) = (G(S), \delta)$ where $G(S)$ is a directed graph with set of vertices $Q(S)$ and set of edges $E(S)$, and $\delta : E(S) \to L_1(S)$ is a labelling map of the edges, which gives the language $L(S)$, that is, the set of words produced by δ along the paths of $G(S)$ is L(S). Moreover, $\mathcal{A}(S)$ satisfies: (i) δ is deterministic: if e, e' are two different edges of $G(S)$ starting from the same vertex then $\delta(e) \neq \delta(e')$, (ii) $G(S)$ has the minimal number of vertices among all graphs satisfying (i). The pair $\mathcal{A}(S)$ is called the minimal Shannon automaton of S. Let us use the short-hand $q \stackrel{u}{\to}$ to state that the word $u \in L(S)$ is the label of a path starting from the vertex $q \in Q(S)$ (define analogously $\stackrel{u}{\to} q$). Since δ is deterministic, $q \stackrel{u}{\to}$ points out the unique path starting from q with label u. We define the SFT of order two, T, over the alphabet $A' = \{(a, q) \in A \times Q(S) / \stackrel{a}{\to} q\}$ by allowing all the words $(a, q)(b, q')$ such that $\stackrel{a}{\to} q \stackrel{b}{\to} q'$. The one-block map $\pi : T \to S$ such that $\pi((a, q)) = a$ is a factor map. The SFT T is called the *minimal Shannon cover* of S. We shall often identify π with T. The minimal Shannon cover of a transitive sofic system is also transitive. If we change the labels $\stackrel{a}{\to} q$ of the minimal Shannon automaton of S to $\stackrel{(a,q)}{\to} q$ we produce a finite automaton describing $L(T)$; it is called the minimal Shannon automaton of T associated to $\mathcal{A}(S)$.

Example. Figure 2.1 shows the minimal Shannon automaton of a sofic system S

and of its minimal Shannon cover T. S is known as the *even system*: the words $w_1 01^{2i+1} 0 w_2$ do not belong to $L(S)$ for all $i \geq 0$, $w_1, w_2 \in \{0,1\}^*$; S is transitive.

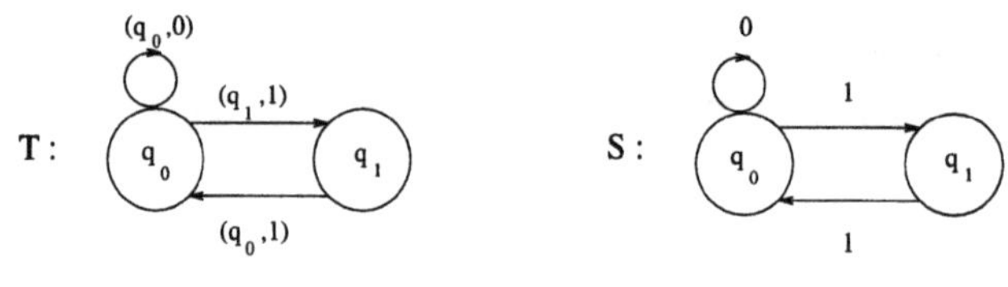

Figure 2.1

The map π satisfies the following properties ([2], [11], [14]):

(i) π is bounded-to-one and right resolving (see below).

(ii) S has a magic word for π: that is, there exists a word $w = w_1....w_l \in L(S)$ and a unique state $q \in Q(S)$ such that $\stackrel{w}{\rightarrow} q$. A word containing a magic word is also magic. This notion of magic word is not the classical one (see [4]), but they are the same when S is transitive. As we only deal with transitive sofic systems we prefer to give this one.

(iii) π is 1-to-1 a.e.: every doubly transitive point of S has a unique preimage by π.

(iv) If S is transitive then $G(S)$ is strongly connected. If S is mixing then T is mixing. If T has a transition length then S has a transition length.

2.2. CLASSES OF MAPS BETWEEN SYMBOLIC FLOWS

Let S and T be two subshifts (not necessarily defined over the same full shift), and let $f : S \rightarrow T$ be a σ–commuting map, that is, $f \circ \sigma = \sigma \circ f$. The continuous σ-commuting maps are characterized in [19] as those maps which are defined by a local rule:

$$\exists\, r \in \mathbb{N},\ \exists\, f^* : L_{2r+1}(S) \rightarrow L_1(T), \text{ such that,}$$

$$\forall\, x \in S,\ \forall\, i \in \mathbb{Z}, \quad f(x)_i = f^*(x_{i-r}...x_i...x_{i+r})$$

Such a map f is called a $(2r+1)$-block map, where r is the radius of f; f^* can be extended in a natural way to $L_m(S)$ for all $m \geq 2r+1$. In the sequel we shall identify f^* with f.

Let $f : S \to T$ be a continuous σ-commuting map of radius r. We say that f is *right closing* if whenever $x = (x_i)_{i \in \mathbb{Z}}$, $y = (y_i)_{i \in \mathbb{Z}} \in S$, $x \neq y$, and there exists an integer n such that for all $i \leq n$ $x_i = y_i$, then $f(x) \neq f(y)$. The composition of right closing maps is also right closing. A particular class of right closing maps are the right M-resolving maps, where M is a positive integer. The map f is *right M-resolving* if whenever $aa_1....a_M \in L_{M+1}(T)$, and $w \in L_{2r+1}(S)$ such that $f(w) = a$, then there is at most one block $b_1....b_M \in L_M(S)$ such that $wb_1....b_M \in L_{2r+M+1}(S)$ and $f(wb_1....b_M) = aa_1....a_M$. When $M = 1$ we say that f is right resolving. Analogously one defines left closing and left M-resolving maps. The map f is *bounded-to-one* if there exists an integer $k \in \mathbb{N}$ such that for any $x \in T$, $card(f^{-1}(x)) \leq k$. Both left closing and right closing block maps whose domain is a transitive SFT are bounded-to-one.

2.3. ONE–DIMENSIONAL CELLULAR AUTOMATA

A (one-dimensional) *cellular automaton* (CA) is a continuous σ-commuting map of a full shift $A^{\mathbb{Z}}$ into itself (in classical symbolic dynamics they are known as endomorphisms of the full shift, see [19]). In other words, a CA is a $(2r+1)$-block map, where r is the radius of the CA (we also call r the neighbourhood radius).

Sometimes it is also interesting to consider CA defined over $A^{\mathbb{N}}$. In this case, if $F : A^{\mathbb{N}} \to A^{\mathbb{N}}$ is a CA, then for any $x \in A^{\mathbb{N}}$, $i \in \mathbb{N}$, $F(x)_i = f(x_i, ..., x_{i+r})$. Where $f : A^{r+1} \to A$ is a given local rule and $r \in \mathbb{N}$ is the radius of F.

2.4. ENTROPY

Let (X, T) be a topological flow and μ be a T-invariant measure. We recall the notions of topological and measure–theoretic entropy.

For defining topological entropy one must introduce some additional notions. Let $\mathcal{R}$ be an open cover of X: we denote by $H(\mathcal{R})$ the real number $\inf\{\log card(\mathcal{R}')\}$, where the inf is taken over the set of finite subcovers $\mathcal{R}'$ of $\mathcal{R}$. Let S be another subcover of X; we say that $\mathcal{R}$ is *finer* than $\mathcal{S}$, and denote this by $\mathcal{S} \leq \mathcal{R}$, if for all $U \in \mathcal{R}$ there exists $V \in \mathcal{S}$ such that $U \subset V$. This implies $H(\mathcal{S}) \leq H(\mathcal{R})$.

Denote by $\mathcal{R} \vee \mathcal{S}$ the cover made up of all the intersections $R \cap S$, where $R \in \mathcal{R}$ and $S \in \mathcal{S}$. The *topological entropy of the cover* $\mathcal{R}$ is the (well defined) nonnegative number

$$h(\mathcal{R},T) = \lim_{n\to\infty} \frac{1}{n} H(\bigvee_{i=0}^{n-1} T^{-i}\mathcal{R}) .$$

Whenever $\mathcal{S} \leq \mathcal{R}$ one has $h(\mathcal{S},T) \leq h(\mathcal{R},T)$.

The *topological entropy* of a flow (X,T) is the number defined by

$$h_{top}(X,T) = \sup h(\mathcal{R},T) ,$$

where the sup is taken over all finite open covers of X.

Here is a list of classical properties of topological entropy we shall use in the sequel; for more details see [13].

(1) Suppose that $(\mathcal{R}_n)_{n\in\mathbb{N}}$ is a generator, i.e. an increasing family of open covers of X with the property that for any other open cover $\mathcal{R}$ of X there is $n \in \mathbb{N}$ such that $\mathcal{R} \leq \mathcal{R}_n$. Then,

$$h_{top}(X,T) = \lim_{n\to\infty} h(\mathcal{R}_n,T) .$$

A consequence is for a subshift S one has

$$h_{top}(S) = \lim_{n\to\infty} \frac{1}{n} \log L_n(S) .$$

(2) If (X',T') is a factor of (X,T) then $h_{top}(X',T') \leq h_{top}(X,T)$.

(3) Let $S \subseteq A^{K_1}$ be a Markov system, then $h_{top}(S) = \log \lambda$, where λ is the maximal eigenvalue of its transition matrix.

We shall only use the definition of measure–theoretic entropy for measures on subshifts. Consider a subshift $X \subseteq A^{K_1}$, where $K_1 = \mathbb{N}$ or $\mathbb{Z}$. The *entropy* of (X,T) *with respect to the invariant measure* μ is,

$$h_\mu(X,T) = -\lim_{n\to\infty} \frac{1}{n} \sum_{w\in L_n(X)} \mu([w]_0) \log \mu([w]_0) . \tag{2.1}$$

In the general case, if the flows (X,T) and (X',T') are conjugate by the map $\pi : X \to X'$, then $h_\mu(X,T) = h_{\pi\mu}(X',T')$.

Both notions of entropy, topological and measure–theoretic, are linked by the classical "variational principle"

$$h_{top}(X,T) = \sup h_\mu(X,T) \ ,$$

where the sup is taken over the set of invariant measures of the flow (X,T). A measure μ for which $h_\mu(X,T) = h_{top}(X,T)$ is said to be a *measure of maximal entropy.*

We shall use the following lemmas extensively.

Lemma 2.2. [2,11] If T is a transitive SFT and $\Lambda \subseteq T$ is a subshift with $h_{top}(\Lambda) = h_{top}(T)$, then $\Lambda = T$.

Lemma 2.3. [11] A factor map $\psi : T \to S$ between the transitive sofic systems T and S is bounded-to-one if and only if $h_{top}(T) = h_{top}(S)$.

3. Dynamical Classifications of Cellular Automata

An aspect of cellular automata that is important for modelling is that certain automata exhibit *self–organizing behaviour*; that is, the property that dissimilar initial conditions tend to similar states under iterations of the automaton. From the dynamical systems point of view this suggests, among other possible interpretations, the presence of attracting sets with large basins. Stephen Wolfram give a classification of CA based on qualitative characterization of long–term bilateral behaviour of random initial conditions [48]:

(C1) all configurations tend to a fixed point for the shift map,
(C2) localized periodic structures appear and persist,
(C3) chaotic behaviour,
(C4) complicated localized structures sometimes long–lived.

There exist several classifications based on different mathematical approaches that try to justify Wolfram's classes. Here we shall present two of them. The first one, due to Hurley, is based on the structure of attractors, and the second one, of Gilman and Kürka uses the notion of equicontinuity.

3.1. CLASSIFICATION WITH RESPECT TO THE ATTRACTORS

Let $T : X \to X$ be a continuous map of the compact metric space X. A closed non empty subset C of X is an *attractor* for T if there exists an open neighbourhood U of C with, (i) $T(\bar{U}) \subseteq U$, and (ii) $C = \cap_{n\geq 0} T^n(U)$. The open set $B(C) = \cup_{n\geq 0} T^{-n}(U)$ is called the *basin of attraction* of C. A closed non empty subset Q of X is a *quasi–attractor* for T if it is a countable intersection of attractors, but itself is not an attractor. We say that an attractor (quasi–attractor) is minimal if it does not contain properly any other attractor (quasi–attractor). It is not difficult to prove that the set of attractors is at most countable.

M. Hurley classify cellular automata as follows,

Theorem 3.1. [21] A cellular automaton $F : A^{\mathbb{Z}} \to A^{\mathbb{Z}}$ satisfies one of the following properties:

(i) there is a unique minimal attractor C for F. In this case C is contained in every attractor of F, C is shift invariant, $B(C)$ is dense and has full measure with respect to the uniform Bernoulli measure.
(ii) there is a unique minimal quasi–attractor C for F. In this case C is contained in every attractor of F and it is shift invariant.
(iii) there is a pair of disjoint attractors. In this case F has uncountably many minimal quasi–attractors.

Proof. We shall give a very simple proof of Petr Kürka [30]. In order to do that recall that the ω–limit set of $Y \subseteq X$ is $\omega(Y) = \cap_{n>0} \overline{\cup_{m>n} T^m(Y)}$. It can be proved that if C_1 and C_2 are attractors with non empty intersection then $\omega(C_1 \cap C_2)$ is also an attractor.

We state the classification for arbitrary dynamical systems $T : X \to X$. Let us suppose that any pair of attractors has a non empty intersection. Let $\mathcal{A} = \{A_0, A_1, ...\}$ be the countable family of attractors associated with T. Define $W_0 = A_0$ and $W_{i+1} = \omega(A_{i+1} \cap W_i)$ for any $i \in \mathbb{N}$, which by hypothesis is a decreasing family of attractors. Therefore the set $W = \cap_{i \in \mathbb{N}} W_i$ is either a minimal attractor or a minimal quasi–attractor. ■

Remark. It is straightforward that the set $\omega(X) = \cap_{i \in \mathbb{N}} T^i(X)$ is an attractor, moreover any other attractor C of T is contained in $\omega(X)$, so you can refine the previous classification by saying:

(1.1) there exists a unique minimal attractor not equal to $\omega(X)$,
(1.2) there exists a unique attractor equal to $\omega(X) \neq X$,
(1.3) X is the unique attractor.

3.2. EQUICONTINUITY CLASSIFICATION

We shall present two classifications of CA based on the "volume" of the set of equicontinuity points. The first one also consider stochastic versions of this property.

Let $F : A^{\mathbb{Z}} \to A^{\mathbb{Z}}$ be a CA. For $n \in \mathbb{N}$ and $x \in A^{\mathbb{Z}}$ define

$$C_n(x) = \{y \in A^{\mathbb{Z}} / \ \forall \ i \in \mathbb{N} \ F^i(x)(-n, n) = F^i(y)(-n, n)\}$$

and for $\epsilon > 0$ put

$$D(x, \epsilon) = \{y \in A^{\mathbb{Z}} / \ \forall \ i \in \mathbb{N} \ d(F^i(x), F^i(y)) < \epsilon\}.$$

Definition 3.2. Fix a Bernoulli measure μ on $A^{\mathbb{Z}}$.

(1) F is equicontinuous at the point x if for any $\epsilon > 0$ there exists $\delta > 0$ such that $d(F^i(x), F^i(y)) < \epsilon$ for all $i \in \mathbb{N}$ and $y \in B_\delta(x) = \{y \in A^{\mathbb{Z}} / d(x, y) < \delta\}$. Denote by $Eq(F)$ the set of equicontinuity points.

(1') F is μ–equicontinuous at the point x if for any $\epsilon > 0$

$$\lim_{n \to \infty} \frac{\mu(C_n(x) \cap D(x, \epsilon))}{\mu(C_n(x))} = 1.$$

(2) F is expansive if there is $\epsilon > 0$ such that for any $y \in X$, $y \neq x$, $d(F^n(x), F^n(y)) \geq \epsilon$ for some $n \in \mathbb{N}$.

(2') F is μ–expansive if there is $\epsilon > 0$ such that for all $x \in A^{\mathbb{Z}}$, $\mu(D(x,\epsilon)) = 0$.

There is an intermediate notion that appears when we study chaotic systems.

(3) F is sensitive (to initial conditions) if:

$$\exists\ \epsilon > 0\ \forall x\ \forall \delta > 0\ \exists y \in B_\delta(x)\ \exists n \in \mathbb{N}\ d(F^n(x), F^n(y)) \geq \epsilon.$$

In this last case it is not difficult to see that $Eq(F) = \emptyset$. ■

Theorem 3.3. [17] Fix a Bernoulli measure μ on $A^{\mathbb{Z}}$. Any cellular automaton $(A^{\mathbb{Z}}, F)$ falls in exactly one of the following cases.

(1) F is equicontinuous at some point $x \in A^{\mathbb{Z}}$,
(2) F is μ–equicontinuous at some point $x \in A^{\mathbb{Z}}$,
(3) F is μ–expansive. ■

Without any measure–theoretical consideration Kürka give the following classification.

Theorem 3.4. [30] Any cellular automaton $(A^{\mathbb{Z}}, F)$ falls in exactly one of the following cases.

(1) $(A^{\mathbb{Z}}, F)$ is equicontinuous.
(2) $\emptyset \neq Eq(F) \neq A^{\mathbb{Z}}$.
(3) $(A^{\mathbb{Z}}, F)$ is sensitive but not expansive.
(4) $(A^{\mathbb{Z}}, F)$ is expansive.

Proof. Since $A^{\mathbb{Z}}$ is compact it is enough to prove that a non sensitive cellular automaton has equicontinuity points.

Let F be a non sensitive CA, then for any $\epsilon > 0$ there is a point $x \in A^{\mathbb{Z}}$ and $\delta > 0$ such that for all $y \in B_\delta(x)$ and any $i \in \mathbb{N}$, $d(F^i(x), F^i(y)) \leq \epsilon$. Therefore there is $m \in \mathbb{N}$ such that if $x, y \in A^{\mathbb{Z}}$ with $x(-m,m) = y(-m,m)$ then $F^i(x)(-1,1) = F^i(y)(-1,1)$ for all $i \in \mathbb{N}$. It follows that the periodic point for the shift $\bar{x}$ of period $2m$ defined by $\bar{x}(-m,m) = x(-m,m)$ is an equicontinuity point. ■

4. Dynamics of onto Cellular Automata

In this section we shall present different results concerning ergodic properties of onto cellular automata. As in the last sections put $K = \mathbb{N}$ or $\mathbb{Z}$ and consider a cellular automaton $F : A^K \to A^K$.

Onto cellular automata are particularly attractive for ergodicians: then $A^{\mathbb{N}}$ or $A^{\mathbb{Z}}$, endowed with the continuous onto self-map F defining the automaton, is a topological flow, in other words a dynamical system "in equilibrium"; another nice feature is that the uniform Bernoulli measure, defined for any cylinder set $[w]_i = \{x \in A^K / x(i, i+n-1) = w\}$, $w \in A^n$, $i \in K$ by

$$\lambda([w]_i) = card(A)^{-n}$$

is always invariant with respect to F. This last fact follows straightforward from the following combinatorial characterization of onto CA proved in [19].

Proposition 4.1. Let $F : A^K \to A^K$ be a CA. The following conditions are equivalent:

(i) F is onto.

(ii) For any $w \in A^*$, $card(\{w' \in A^{|w|+r(K)} / F(w') = w\}) = card(A)^{r(K)}$, where $r(\mathbb{N}) = r$ and $r(\mathbb{Z}) = 2r$ (recall that r is the radius of F). ■

Thus the measure–theoretic flow $\mathcal{X} = (A^K, F, \lambda)$ can be analyzed under the scope of ergodic theory and topological dynamics. One of the fundamental problems in ergodic theory is to classify measure–theoretic dynamical systems. In this direction it appears that properties like ergodicity, or the concept of entropy play a fundamental role. Interesting classes of dynamical systems, like K–automorphisms or Bernoulli systems have proved to possess strong ergodic properties. We are going here to prove some of these properties for some classes of cellular automata.

Every onto cellular automaton F has associated a family of symbolic factors (or subshifts) $\mathcal{S} = \{S_F(i,j) \; ; \; i, j \in K, \; i \le j\}$, where

$$S_F(i,j) = \{y \in (A^{j-i+1})^{\mathbb{N}} \; / \; \exists \; x \in A^K, \; F^l(x)(i,j) = y_l, \; l \in \mathbb{N}\}.$$

In other words, we code the orbits of points with respect to the partitions of clopen sets $\alpha(i,j) = \{[w]_0 \ / \ w \in A^{j-i+1}\}$. One element of $S(i,j)$ represents the *temporal configuration* generated by the dynamics of x between the coordinates i and j. The subshifts $S_F(i,j)$ play a fundamental role in the dynamics of the cellular automata; in particular, they give the topological entropy (see for example [5]), mixing properties can also be deduced from those subshifts, etc.

The subshift $S_F(i,j)$ is a topological factor of the cellular automaton dynamical system $\mathcal{X}$, thus there is a continuous and onto map

$$\pi_F(i,j) : A^K \to S_F(i,j)$$

such that $\sigma \circ \pi_F(i,j) = \pi_F(i,j) \circ F$. If there are $i,j \in K$ such that $\pi_F(i,j)$ is also one to one (so it is bijective) then we say that F is expansive. This definition is equivalent with the classical notion of "positive expansiveness" (proposition 10 of [5]): a topological flow (X,T) is positively expansive (expansive for short) if it exists a positive constant ϵ such that for every x, $y \in X$, $x \neq y$, we have $d(F^l(x), F^l(y)) > \epsilon$ for some $l \in \mathbb{N}$. Whenever a cellular automaton is expansive its dynamics depends only of the subshift $S_F(0,2r-1)$ if $K = \mathbb{Z}$, or $S_F(0,r-1)$ if $K = \mathbb{N}$. We say that last subshifts are the canonical factors of $\mathcal{X}$.

If a cellular automaton is not onto, we can always study the sets $S_F(i,j)$, but they are not more subshifts. In this case, a computer science approach consists of consider the languages $L(S_F(i,j))$ which are always context–sensitive languages.

Example. The CA $F : \{0,1\}^{\mathbb{Z}} \to \{0,1\}^{\mathbb{Z}}$ defined for any $x \in \{0,1\}^{\mathbb{Z}}$ and $i \in \mathbb{Z}$ by $F(x)_i = 1$ if $x_{i+1} = x_{i+2} = 1$ and $F(x)_i = 0$ otherwise, has a context–sensitive, non context-free, canonical factor. ■

Before going to the results let us describe some families of onto cellular automata. We shall suppose that the local rule of the cellular automaton F is defined by two parameters $r, l \in K$, $l \leq r$, such that for any $x \in A^K$, $i \in K$, $F(x)_i = f(x_{i+l}, ..., x_{i+r})$.

Example 1. *Toggle automata.* A cellular automaton $F : A^K \to A^K$ is a right–toggle automaton if for every $w \in A^{r-l}$ there is a permutation π_w of A such that $f(wa) = \pi_w(a)$ for all $a \in A$. In a similar way we define left–toggle automata.

Example 2. *A generalisation.* Suppose here that F is one sided (i.e. $K = \mathbb{N}$, $l = 0$ and $r \geq 0$). We construct a family of automata that are permutative in a weaker sense: a permutation is attached to every word in a finite set, but here the set is no longer A^r. Fix an integer $r > 2$ and consider a finite language $\mathcal{C} \subseteq A^r \cup A^{r-1}$ such that:

(1) $\mathcal{C}$ is a complete prefix code: if $c_0...c_k \in \mathcal{C}$, then for all $0 \leq i < k$, $c_0...c_i \notin \mathcal{C}$, and if $x \in A^{\mathbb{N}}$, there exist (unique) $c \in \mathcal{C}$ and $y \in A^{\mathbb{N}}$ such that $x = cy$.

(2) If $c_0...c_{r-1} \in \mathcal{C}$, then for all $\alpha \in A$, $c_1...c_{r-2}\alpha \in \mathcal{C}$.

For instance, $\mathcal{C} = \{00, 01, 11, 100, 101\}$ satisfies the last two conditions with $r = 3$ and $A = \{0, 1\}$.

Now, we associate to each word $w \in \mathcal{C}$ a permutation $\pi_w : A \to A$ with the following property: let $w = a_0a_1...a_{r-2}\alpha \in \mathcal{C}$, $w' = a_0a_1...a_{r-2}\alpha' \in \mathcal{C}$, where $\alpha \neq \alpha'$; if for b, $b' \in A$, $\pi_w(b) = \pi_{w'}(b')$ then $\pi_{(a_1...a_{r-2}\alpha)}(b) \neq \pi_{(a_1...a_{r-2}\alpha')}(b')$. This choice is possible: first fix arbitrarily the permutations associated with the words of length $r - 1$, and then choose the missing ones in such a way that the property holds.

Finally, define the cellular automaton $F_{\mathcal{C}} : A^{\mathbb{N}} \to A^{\mathbb{N}}$ such that for all $x \in A^{\mathbb{N}}$ and $i \in \mathbb{N}$, $F_{\mathcal{C}}(x)_i = \pi_w(x_{i+|w|})$, where w is the unique prefix of $x(i, i + r)$ in $\mathcal{C}$. Evidently when $\mathcal{C} \cap A^r = \emptyset$ we have defined a toggle automaton.

For $F_{\mathcal{C}}$ the following holds: for all w, $w' \in A^{r-1}$ there is $w'' \in A^r$ such that $F_{\mathcal{C}}(ww'') = w'$, and if w_1'' and w_2'' are words in A^r for which last equality holds then they have the same first letter. It follows that $F_{\mathcal{C}}$ is expansive and that $S_{F_{\mathcal{C}}}(0, r - 1)$ is conjugate (but not equal) to $((A^{r-1})^{\mathbb{N}}, \sigma)$.

Example 3. *Coven's examples.* Put $A = \{0, 1\}$ and consider the local rule $f : A^{r+1} \to A$ such that for any $x \in A^{\mathbb{Z}}$, $F(x)_0 = f(x_0x_1....x_r) = x_0 + \prod_{i=1}^{r}(x_i + b_i + 1)$, where $B = b_1...b_r$ is a given word in A^* (the multiplication and addition in last formula are $\bmod_2$).

Whenever B is an aperiodic block, that is, there is no overlapping of B with itself, we can describe completely the dynamics of Coven's CA. In [7], we proved that (i) $\forall i, j \in \mathbb{Z}$, $S_F(i, j)$ are sofic systems but not S.F.T., (ii) they contain equicontinuous points without being equicontinuous, (iii) they are chain transitive, but not topologically transitive, and (iv) their entropy is $\log 2$.

Example 4. *Multiplication automata.* For $k \in \mathbb{N}$ put $A_k = \{0, ..., k-1\}$. First fix a pair $(k_1, k_2) \in \mathbb{N}^2$ and define $p = k_1\ k_2$. Each configuration of $A_p^{\mathbb{N}}$ is the expansion in base p of a real number in the interval $[0,1]$: it is proved in [6] that for all $k \in \mathbb{N}$ such that k divides an integer power of p there exists a CA, whose alphabet is A_p, which represents the multiplication by k in base p. This follows from the fact that, under the last condition, the algorithm of multiplication by an integer in base p only depends of a finite number of carries.

In particular, consider the case $k = k_1$. Each integer $a \in A_p$ can be written $a = f_a k_2 + l_a$ where $f_a \in A_{k_1}$ and $l_a \in A_{k_2}$, or also $a = \bar{f}_a k_1 + \bar{l}_a$ with $\bar{f}_a \in A_{k_2}$ and $\bar{l}_a \in A_{k_1}$. It is not difficult to check that the set of carries obtained in the multiplication by k_1 in base p is equal to A_{k_1}. Let $a \in A_p$: independently of the value of $r \in A_{k_1}$, the carry obtained when we multiply a by k_1 and add r is always f_a; in fact, $k_1 a + r = k_1 k_2 f_a + l_a k_1 + r = f_a p + (l_a k_1 + r)$, where $l_a k_1 + r \leq p - 1$. The onto cellular automaton F_{k_1,k_2} representing the multiplication by k_1 in base p is then defined by

$$F_{k_1,k_2}(x)_i = k_1\ l_{x_i}\ +\ f_{x_{i+1}},\ x \in A_p^{\mathbb{N}},\ i \in \mathbb{N}.$$

It is clear that $F_{k_1,k_2}(x)_i \in A_p$, and that its radius is 1. It was proved in [9] that F_{k_1,k_2} is expansive if and only if k_2 divides a positive power of k_1.

———— o ————

There is a natural question that arise when we consider ergodic properties of CA: What about the ergodicity of the flow $\mathcal{X}$? The class defined in example 3 provides with some examples of CA which are not ergodic with respect to λ. In particular, take the cellular automaton defined by the local map: $f(x_0, x_1, x_2) = x_0 + x_1(x_2 + 1)$; it is not topologically transitive and therefore by proposition 2.1 it cannot be ergodic. Indeed, a simple computation shows that $S_F(0,1)$ is topologically conjugate to the disjoint union of one–sided full shifts $\mathcal{Y} = \{0,1\}^{\mathbb{N}} \cup \{1,3\}^{\mathbb{N}} \cup \{2,3\}^{\mathbb{N}}$.

One of the first articles where it was studied the ergodicity of the flows $(A^{\mathbb{Z}}, F, \lambda)$ is [46]. In this work Shirvani and Rogers consider two families of cellular automata: (1) those ones where the local rules depend of strictly positive or negative coordinates, i.e. $0 < l \leq r$ or $l \leq r < 0$, and (2) the class of right–toggle automata (and left–toggle automata) where $l \leq 0 < r$ (respectively $l < 0 \leq r$). They proved that cellular automata in both classes are strongly mixing with respect to the uniform measure λ; in particular they are ergodic.

After, in [43] Shereshevsky find strong ergodic properties, such as the "K" property or the conjugation with Bernoulli systems, of the natural extensions of cellular automata belonging to the following particular classes:

(C1) right–toggle automata with $0 \leq l < r$, or left–toggle automata with $l < r \leq 0$,
(C2) right–toggle automata with $0 < r$, or left–toggle automata with $l < 0$,
(C3) surjective cellular automata with $0 < l \leq r$ or $l \leq r < 0$.

Let us recall the notions of *natural extension* and *K–automorphism*. The *natural extension* of a flow (X, T) is the flow $(\tilde{X}, \tilde{T})$, where $\tilde{X} = \{\tilde{x} = (x^{(0)}, x^{(1)}, ...) \in X^{\mathbb{N}} /$ for all $i \geq 0$, $T(x^{(i+1)}) = x^{(i)}\}$ and $\tilde{T}((x^{(0)}, x^{(1)}, ...)) = (T(x^{(0)}), x^{(0)}, x^{(1)}, ...)$. $\tilde{T}$ is obviously an automorphism of $\tilde{X}$. The flow (X, T) is a factor of $(\tilde{X}, \tilde{T})$ for the projection with respect to the first coordinate. Let μ be a T-invariant measure, and $\mathcal{B}$ be the Borel sigma-algebra on X. We define $\tilde{\mu}$ over the sigma-algebra $\tilde{\mathcal{B}}$, generated by the family $C^{(i)} = \{\tilde{x} \in \tilde{X} / x^{(i)} \in C\}$, $C \in \mathcal{B}$, by: $\tilde{\mu}(C^{(i)}) = \mu(C)$. A flow (X, T) is a *K-system* with respect to the invariant measure μ if it is bijective and there exists a sigma-algebra $\mathcal{A} \subseteq \mathcal{B}$ such that

$$\mathcal{A} \subseteq T(\mathcal{A}), \quad \bigvee_{i=0}^{\infty} T^{i}(\mathcal{A}) = \mathcal{B}, \quad \bigcap_{i=0}^{\infty} T^{-i}(\mathcal{A}) = \mathcal{N},$$

where $\mathcal{N}$ is the trivial σ-algebra of X.

Theorem 4.2. Let $F : A^{\mathbb{Z}} \to A^{\mathbb{Z}}$ be a surjective cellular automaton.

(1) If F belongs to (C1), (C2) or (C3) then it is strongly mixing of all orders with respect to λ.
(2) If F belongs to class (C1) then its natural extension is a Bernoulli automorphism.

(3) If F belongs to class (C3) then its natural extension is a K–automorphism.

Proof. For the proof you can see [43]. ∎

From now on in this section we shall only consider one–sided onto cellular automata. Motivated readers can try to extend the results presented here to the two–sided case.

As was shown by Shereshevsky, one–sided right–toggle cellular automata present the following dynamical properties: the canonical factor is a full shift, and they are *expansive* in the sense that $(A^{\mathbb{N}}, F)$ is conjugate to $(S_F(0, r-1), \sigma)$. Now we present results dealing with the symbolic dynamics of the family of CA that, as topological flows, are conjugate to subshifts - in other words, the abstract family of expansive one-sided cellular automata - and their ergodic behaviour when $A^{\mathbb{N}}$ is endowed with the uniform measure. It turns out that the expansiveness assumption has many strong symbolic and measure-theoretic consequences [9].

We have to mention that from theorem 1.3 in [18] it can be deduced straightforward that transitive cellular automata (in particular they are onto) are sensitive to initial conditions, so expansive CA is a natural class (of possible transitive) sensitive systems to study.

Put $S_F = S_F(0, r-1)$ and $\pi_F = \pi_F(0, r-1)$. For w, $w' \in A^* \cup A^{\mathbb{N}}$, we say that w' is a *successor* of w with respect to F, and denote this by $w \xrightarrow{F} w'$, if there are $x_w = wy_w$ and $x_{w'} = w'y_{w'}$ in $A^{\mathbb{N}}$ such that $F(x_w) = x_{w'}$.

The following lemma reminds one of the characterizations of shift-commuting or right resolving maps in symbolic dynamics. It relies on a simple compactness argument.

Lemma 4.3. Let $F : A^{\mathbb{N}} \to A^{\mathbb{N}}$ be an expansive CA. There are a positive integer ℓ, and a map $\bar{f} : L_{\ell+1}(S_F) \to A^r$ such that for all $w = w_0 ... w_\ell \in L_{\ell+1}(S_F)$ and $x \in A^{\mathbb{N}}$ such that $F^i(x)(0, r-1) = w_i$, $i = 0, ..., \ell$, one has $x(r, 2r-1) = \bar{f}(w)$. ∎

Lemma 4.3 expresses the existence of the *transposed flow* $(S_F, \bar{F})$ of an expansive CA F, where $\bar{F} = \pi_F \circ \sigma^r \circ \pi_F^{-1}$ is also surjective, and whose local rule $\bar{f}$ is defined by: if y, $y' \in S_F$ and $\bar{F}(y) = y'$, then $y'_i = \bar{f}(y(i, i+\ell))$ for any $i \in \mathbb{N}$. It is not difficult to check that $(S_F, \bar{F})$ is also expansive.

Remark. Last lemma implies in particular that for each $a_0 \in A^r$ and $b_1...b_\ell \in (A^r)^*$ such that $a_0 \xrightarrow{F} b_1...b_\ell$ there is a unique $a_1 \in A^r$ for which $a_0a_1 \xrightarrow{F} b_1...b_\ell$. This proposition is also true for one-sided right closing maps.

An important case is when $r = 1$; we then say F is *of type (E1)*. Notice that F is always conjugate to a CA $F' : (A^r)^{\mathbb{N}} \to (A^r)^{\mathbb{N}}$ of type (E1). Indeed, let us define the local rule $f' : (A^r)^2 \to A^r$ associated to F' by $f'(w_1, w_2) = F(w_1w_2)$ for $w_1,\ w_2 \in A^r$; as usual we identify F' and f'. The map $\phi : A^{\mathbb{N}} \to (A^r)^{\mathbb{N}}$ such that $\phi(x)_i = x(ir, ir + r - 1)$, for every $x \in A^{\mathbb{N}}$ and $i \in \mathbb{N}$, is a conjugacy between $(A^{\mathbb{N}}, F)$ and $((A^r)^{\mathbb{N}}, F')$. Naturally F' is also expansive.

Theorem 4.4. Let F be an expansive CA on $A^{\mathbb{N}}$; then S_F is a subshift of finite type.

Proof. We shall prove that if F is expansive it is an open map (this is an idea of P. Kůrka [30]). This implies that the conjugate map $\sigma : S_F \to S_F$ is also open. Parry's result that one-sided subshifts for which the shift is an open map are exactly the SFT [42] will end the proof.

By recoding we can suppose that F is of type (E1). It is enough to prove the following assertion:

Claim: Suppose that $a_0 \in A$ and $b_1...b_{\ell-1} \in A^{\ell-1}$ are such that $a_0 \xrightarrow{F} b_1...b_{\ell-1}$. Then for all $\beta \in A$ there is a unique $\alpha \in A$ such that $a_0\alpha \xrightarrow{F} b_1...b_{\ell-1}\beta$.

Suppose that the claim is not true, then there exist $a_0, \beta \in A$ and $b_1...b_{\ell-1} \in A^*$ with $a_0 \xrightarrow{F} b_1...b_{\ell-1}$ but $b_1...b_{\ell-1}\beta$ not being a successor of a_0. Put $A = \{a_0, a_1, ..., a_{|A|-1}\}$.

Since F is onto there is $\alpha \in A$ such that the word $b_1...b_{\ell-1}\beta$ is one of their successors. We can suppose without loss of generality that $a_1 \xrightarrow{F} b_1...b_{\ell-1}\beta$ and take $v_1...v_\ell \in A^\ell$ such that $F(a_1v_1...v_\ell) = b_1...b_{\ell-1}\beta$. It is straightforward that

$$a_1v_1...v_\ell a_0 \xrightarrow{F} b_1...b_{\ell-1}\beta\delta b_1...b_{\ell-1},$$

for some $\delta \in A$, but $w_1 = b_1...b_{\ell-1}\beta\delta b_1...b_{\ell-1}\beta$ is not a successor of a_1 because by lemma 4.3 any preimage of w_1 has the suffix $a_1v_1...v_\ell a_0$. Therefore, w_1 is neither a successor of a_0 nor of a_1.

We can repeat the same construction to extend the word w_1 to the right until obtain a word $w \in A^*$ without preimage in A^*, which is a contradiction. ■

Put $N = \max(\text{order of } S_F, \ell)$.

Lemma 4.5. Let $F : A^{\mathbb{N}} \to A^{\mathbb{N}}$ be a CA of type (E1). Suppose that $a_0...a_{N-1} \in A^N$ and $b_1...b_{N-1} \in A^{N-1}$ are such that $a_0 \xrightarrow{F} b_1...b_{N-1}$ and for all $i \in \{0,...,N-2\}$ $F^i(b_1...b_{N-1})(0) = a_{i+1}$. Then, for all $b'_1...b'_{N-1} \in A^{N-1}$ such that $F^i(b'_1...b'_{N-1})(0) = a_{i+1}$, $i \in \{0,...,N-2\}$, one has $a_0 \xrightarrow{F} b'_1...b'_{N-1}$.

Proof. Let $b'_1...b'_{N-1} \in A^{N-1}$ be such that $F^i(b'_1...b'_{N-1})(0) = a_{i+1}$, $i \in \{0,...,N-2\}$, and consider the configuration $z' = b'_1...b'_{N-1}z'_1z'_2.... \in A^{\mathbb{N}}$, where $(z'_i)_{i\in\mathbb{N}}$ is arbitrary. Since $\pi_F(z')_i = a_{i+1}$ for all $i \in \{0,...,N-2\}$ and the order of S_F is N, then $a_0\pi_F(z') \in S_F$. It follows that $F(a_0x') = z'$ for at least one $x' \in A^{\mathbb{N}}$. This last fact completes the proof of the lemma. ■

Let $F : A^{\mathbb{N}} \to A^{\mathbb{N}}$ be expansive. For $a_1...a_{N-1} \in L_{N-1}(S_F)$ define

$$k_1(a_1...a_{N-1}) = card(\{\ a_0 \in A^r / a_0a_1...a_{N-1} \in L_N(S_F)\ \})\ . \qquad (4.1)$$

By Lemma 4.5, using the conjugacy of F with a CA of type (E1) described at the beginning of this section, one gets

$$k_1(a_1...a_{N-1}) = card(\{\ a_0 \in A^r / a_0 \xrightarrow{F} b_1...b_{N-1}\ \})\ , \qquad (4.2)$$

for any choice of the words $b_1,...,b_{N-1} \in A^r$ such that $F^i(b_1...b_{N-1})(0, r-1) = a_{i+1}$, $i \in \{0,...,N-2\}$. On the other hand, for $a_0 \in A^r$ and $b_1,...,b_{N-1} \in A^r$ such that $a_0 \xrightarrow{F} b_1...b_{N-1}$ define

$$k_2(a_0\ ; b_1...b_{N-1}) = card(\{\ w \in A^{r(N-1)} / F(a_0w) = b_1...b_{N-1}\ \})\ . \qquad (4.3)$$

Lemma 4.6. Let $F : A^{\mathbb{N}} \to A^{\mathbb{N}}$ be expansive. The map $k_2(\cdot\ ;\ \cdot)$ is constant over the pairs of words $(a_0\ ; b_1...b_{N-1}) \in A^r \times (A^r)^{N-1}$ such that $a_0 \xrightarrow{F} b_1...b_{N-1}$.

Proof. See [9] ■

The constant defined above will be denoted by k_2. The following proposition is direct from lemma 4.6 and proposition 4.1.

Proposition 4.7. Let $F : A^{\mathbb{N}} \to A^{\mathbb{N}}$ be expansive. Then $k_1(\cdot) : L_{N-1}(S_F) \to \mathbb{N}$ is a constant denoted by k_1 and $card(A)^r = k_1\ k_2$. ■

We can state the main theorem of this section.

Theorem 4.8. [9] Let $F : A^{\mathbb{N}} \to A^{\mathbb{N}}$ be an expansive cellular automaton. Then,

(1) (S_F, σ) is a mixing subshift of finite type,
(2) F is topologically mixing,
(3) $h_{top}(F) = \log k_1$ where $k_1 \in \mathbb{N}$,
(4) λ is a measure of maximal entropy for F,
(5) the natural extension of the flow $\mathcal{X} = (A^{\mathbb{N}}, F, \lambda)$ is a Bernoulli system.

Proof. By recoding we can suppose that F is of type (E1).

(1,2) Since S_F and $(A^{\mathbb{N}}, F)$ are conjugate and S_F is of finite type (theorem 4.4) we only have to prove that S_F is mixing. To show S_F is mixing it is enough to find a constant $t_0 \in \mathbb{N}$ such that for all $t \geq t_0$ and $w = w_0 ... w_{N-2}$, $\tilde{w} = \tilde{w}_0 ... \tilde{w}_{N-2}$ $\in L_{N-1}(S_F)$, there exists $u \in A^t$ satisfying $wu\tilde{w} \in L(S_F)$.

First observe that there is a constant m_0, that we choose larger than $3N$, such that if $x, y \in A^{\mathbb{N}}$ satisfy $F^i(x)(0) = F^i(y)(0)$, $i = 0, ..., m_0 - 1$, then

$$F^0(x)(N-1).....F^{N-2}(x)(N-1) = F^0(y)(N-1).....F^{N-2}(y)(N-1);$$

this is proved by applying Lemma 4.3 repeatedly. In other words this means that the first m_0 coordinates of $\pi_F(x)$ completely determine the first $N-1$ coordinates of $\pi_F(\sigma^{N-1}x)$.

Fix a point $x \in A^{\mathbb{N}}$ in such a way that w is a prefix of $(F^i(x)(N-1))_{i\in\mathbb{N}}$, that is, $w = F^0(x)(N-1).....F^{N-2}(x)(N-1)$. One can do this because $w \in L(S_F)$. Put $v = F^0(x)(0).....F^{m_0-1}(x)(0) = \tilde{v}\tilde{w}'$, with $|\tilde{w}'| = N-1$. Of course $v \in L(S_F)$ too. Now we forget about x and only keep v in mind.

Now consider another point $\tilde{x} \in A^{\mathbb{N}}$ such that $\tilde{w}' = F^0(\tilde{x})(0).....F^{N-2}(\tilde{x})(0)$ and $\tilde{w} = F^0(\tilde{x})(N-1).....F^{N-2}(\tilde{x})(N-1)$. To construct it, since $\tilde{w}$ and $\tilde{w}'$ belong to $L_{N-1}(S_F)$, we only have to concatenate a preimage of $\tilde{w}'$ of length $N-1$ with a preimage of length $N-1$ of $\tilde{w}$, then arbitrarily complete this word to the right into a configuration. Let $\pi_F(\tilde{x})$ be the unique element of S_F associated to $\tilde{x}$.

Since the order of S_F is N, then $\tilde{v}\pi_F(\tilde{x}) \in S_F$. Since F is expansive there is $y \in A^{\mathbb{N}}$ such that $\pi_F(y) = \tilde{v}\pi_F(\tilde{x})$ and $F^{m_0-N+1}(y) = \tilde{x}$. But since v is a prefix of length m_0 of $\pi_F(y)$, then by the observation above w is a prefix of $\pi_F(\sigma^{N-1}y)$; therefore $F^0(y)(N-1).....F^{m_0-1}(y)(N-1) = wu\tilde{w} \in L(S_F)$.

Setting $t_0 = m_0 - 2(N-1)$ and repeating the same construction for $m > m_0$ completes the proof of items (1) and (2).

(3) This follows immediately from propositions 4.7 and 4.1.

(4) Since $(A^{\mathbb{N}}, F, \lambda)$ is conjugate to $(S_F, \sigma, \lambda_F = \pi_F(\lambda))$, we have to compute

$$h_{\lambda_F}(S_F) = \lim_{n\to\infty} -\frac{1}{n} \sum_{w\in L_n(S_F)} \lambda_F([w]_0)\log\lambda_F([w]_0) \ .$$

Fix $n \geq N-1$. For $w = w_0...w_{n-1} \in L_n(S_F)$, let us show first that $\lambda_F([w]_0)$ depends only on n and the last $N-1$ letters of w. By definition,

$$\begin{aligned}\lambda_F([w]_0) &= \lambda(\pi_F^{-1}([w]_0)) \\ &= \frac{1}{card(A)^n}\, card(\{w' \in A^n / F^i(w')(0) = w_i,\ i = 0, ..., n-1\}) \ .\end{aligned}$$

Denote by $u = u_0...u_{N-2}$ the suffix of length $N-1$ of w and by $k^*(u)$ the cardinality of the set

$$S(u) = \{v \in A^{N-1} /\ F^i(v)(0) = u_i,\ i = 0, ..., N-2\} \ .$$

By Lemma 4.6 and proposition 4.7, for each $v \in S(u)$ there exist k_2^{n-N+1} words $w' \in A^n$ such that $F^i(w')(0) = w_i$, $i = 0, ..., n-1$, and $F^{n-N+1}(w')(0, N-2) = v$. Therefore

$$card(\{w' \in A^n / F^i(w')(0) = w_i,\ i = 0, ..., n-1\}) = k^*(u)\ k_2^{n-N+1} \ ,$$

and

$$\lambda_F([w]_0) = \frac{1}{card(A)^n}\; k^*(u)\; k_2^{n-N+1}\;, \tag{4.4}$$

which by Proposition 4.7 can be rewritten

$$\lambda_F([w]_0) = \frac{1}{k_1^n}\; C\; k^*(u),$$

where C does not depend on n and w. Noticing that the last equality implies

$$k_1^n = \sum_{w \in L_n(S_F)} C\; k^*(u),$$

by a simple computation one obtains

$$h_{\lambda_F}(S_F) = \log k_1.$$

(5) By the second part of this theorem, S_F is a mixing SFT, then by Parry's result ([41], or theorem (19.14) in [13]), it bears a unique measure of maximal entropy μ_{max} called the Parry measure; this measure must coincide with λ_F. Since λ_F is the Parry measure of a mixing SFT, by [13] (17.15, corollary 4) it is a Bernoulli system of entropy $\log k_1$. Now, by the Ornstein theorem (see [13] (12.10)) it is isomorphic to the full shift with k_1 elements endowed with the uniform measure. ■

To finish this section let us mention some results concerning the two-sided case, that is, when the CA domain is $A^{\mathbb{Z}}$. Here, we can also consider positively expansive CA, in the sense that $(A^{\mathbb{Z}}, F)$ is topologically conjugate to one-sided subshifts. The dynamics of this class of CA was completely described by Nasu [40].

Theorem 4.9. ([40]) Let $F : A^{\mathbb{Z}} \to A^{\mathbb{Z}}$ be a positively expansive CA. Then $(A^{\mathbb{Z}}, F)$ is topologically conjugate to a one-sided full shift. Measure-theoretical, the uniform Bernoulli measure λ is the maximal measure of the flow.

Proof. Let $F : A^{\mathbb{Z}} \to A^{\mathbb{Z}}$ be a positively expansive CA. Then $(A^{\mathbb{Z}}, F)$ is conjugate to the subshift $S_F(0, 2r-1)$, where r is the neighbourhood radius of F (the proof uses the same arguments that we used in the one-sided case). Moreover, by recoding we can suppose that $r = 1$, and thus the CA is conjugate to $S_F = S_F(0,1)$. By compactness, we deduce that there is a block map $F^* : S_F \to S_F$ with the properties: (i) F^* is an automorphism of S_F, (ii) (S_F, F^*) is topologically conjugate to $(A^{\mathbb{Z}}, \sigma)$. The map F^* is called the transposed flow associated to the positively expansive CA F.

As in the one-sided case S_F is a mixing subshift of finite type. Here we will not give the complete proof because it can be easily recovered by the reader from the arguments given in the proofs of Theorem 4.8 and Theorem 4.4. The main point is that it exists $N \in \mathbb{N}$ such that for any $a, c, b_0, ..., b_{N-1}, d_0, ..., d_{N-1} \in A$ with $ca \overset{F}{\to} d_{N-1}...d_0b_0...b_{N-1}$, and β, $\delta \in A$ there are unique $\alpha, \gamma \in A$ such that $\gamma ca\alpha \overset{F}{\to} \delta d_{N-1}...d_0b_0...b_{N-1}\beta$ (here, $ca \overset{F}{\to} d_{N-1}...d_0b_0...b_{N-1}$ means that there are words $w, w' \in A^N$ such that $F(wcaw') = d_{N-1}...d_0b_0...b_{N-1}$). This property implies that F is an open map and then the shift map $\sigma : S_F \to S_F$ is also open. We conclude that S_F is a subshift of finite type.

By standard recoding we can prove that $F^* : S_F \to S_F$ is topologically conjugate to an automorphism $G : T \to T$, where G has radius $\underline{1}$ and T is a one-side topologically mixing Markov shift. Moreover, we can suppose that T is described by a graph $\Gamma = (V, E)$ (V is the set of vertex and E is the set of edges) in such a way that

$$T = \{x \in E^{\mathbb{N}} / \forall i \in \mathbb{N}, \text{ the terminal vertex of } x_i = \text{ initial vertex of } x_{i+1}\}.$$

Also, the flow (T, G) is topologically conjugate to $(A^{\mathbb{Z}}, \sigma)$. Since G is an automorphism, then it is left resolving, that is, for any $e, f, \bar{f} \in E$ such that $e \overset{G}{\to} f$ and $\bar{f}f \in L_2(T)$ there is a unique $\bar{e} \in E$ such that $\bar{e}e \overset{G}{\to} \bar{f}f$.

First we will prove that T is conjugate to a one-sided full shift, for that, let us begin with some claims.

Claim 1: If ee_1 and ee_2 are path of Γ and $f_1 = G(ee_1)$, $f_2 = G(ee_2)$, then the initial vertex of f_1 and f_2 coincide.

Proof. Take $b \in E$ such that $be \in L_2(T)$, then both bee_1 and bee_2 are path of the graph Γ. Since $G(bee_1) = G(be)G(ee_1)$ and $G(bee_2) = G(be)G(ee_2)$, the terminal vertex of $G(be)$ must be the initial vertex of f_1 and f_2. ■

The claim shows that the map $s : E \to V$ that associates to each $e \in E$ the initial vertex of $G(ef)$ for any $f \in E$ such that $ef \in L_2(T)$, is well defined.

Claim 2: Let $e, e' \in E$ be edges starting from the same vertex $u \in V$. Then for any $v \in V$ the number of edges in Γ from v to $s(e)$ and from v to $s(e')$ is the same.

Proof. Let $g \in E$ be an edge with terminal vertex $s(e)$. Then there are $\bar{g}, \bar{e} \in E$ such that $G(e\bar{e}) = \bar{g}$ and by claim 1 the initial vertex of $\bar{g}$ is $s(e)$. Since G is left resolving there is $f \in E$, with terminal vertex u such that $G(fe) = g$, which implies that the initial vertex of g is $s(f)$. On the other hand, if f_1, f_2 are different edges with terminal vertex u then $G(f_1e) \neq G(f_2e)$ (G is an automorphism). Thus, there is a one-to-one correspondance between the edges with terminal vertex u and the vertex with terminal vertex $s(e)$. Analogously, there is a bijection between the edges arriving to u and those with terminal vertex $s(e')$. Since any edge coming to $s(e)$ or $s(e')$ has initial vertex $s(f)$ for some edge f with terminal vertex u we conclude the claim. ■

If the graph Γ is "column reduced" (that is, for any $u, v \in V$ there is $\bar{v} \in V$ such that the number of edges from $\bar{v}$ to u and the number of edges from $\bar{v}$ to v are different), claim 2 implies that $s(e) = s(e')$ whenever e and e' starts from the same vertex. In this case you can define a map $\varphi : V \to V$ such that $\varphi(v) = s(e)$ where e is any edge with initial vertex v. Since G is topologically mixing you can order the vertex in V in such a way that $V = \{v_0, ..., v_{\ell-1}\}$ and $\varphi(v_0) = v_1$, $\varphi(v_1) = v_2, ..., \varphi(v_{\ell-1}) = v_0$. But, again using the mixing property, ℓ must be equal to 1. This means that $V = \{v\}$ and T is a full shift, proving the assertion of the theorem.

But in general Γ is not column reduced. In this case we know that there is a 1-block conjugation map $\pi : T \to T'$, where T' is given by a graph Γ' that is column reduced. In particular π is left-resolving. Therefore, we can define (in the same way we did for the map G) the maps $s_\pi : E \to V'$ and $s_{G_\pi} : E \to V'$, where

$G_\pi = \pi \circ G$, such that for $e \in E$, $s_\pi(e)$ is the initial vertex in V' of $\pi(e)$ and $s_{G_\pi}(e)$ is the initial vertex in V' of $G_\pi(ef)$ for any $f \in E$ such that $ef \in L_2(T)$. Since Γ' is column reduced we have that $s_\pi(e) = s_\pi(e')$ and $s_{G_\pi}(e) = s_{G_\pi}(e')$ if they have the same initial vertex in V (use claim 2 for π and G_π instead of G).

Claim 3: If $s_\pi(e) = s_\pi(e')$ then $s_{G_\pi}(e) = s_{G_\pi}(e')$.

Proof. Let us suppose that $s_{G_\pi}(e) \neq s_{G_\pi}(e')$. Since Γ' is column reduced, then e and e' cannot have the same initial vertex. Furthermore, $s_\pi(e)$ (that is equal to $s_\pi(e')$) has an incoming edge g such that:

(i) $\exists \bar{u}_1, \bar{u}_2 \in V$, $\bar{u}_1 \neq \bar{u}_2$ and edges $g_e, g_{e'} \in E$ such that $g_e e \in L_2(T)$, $g_{e'} e' \in L_2(T)$ and $\pi(g_e) = g$, $\pi(g_{e'}) = e_0$

(ii) $s_{G_\pi}(g_e) \neq s_{G_\pi}(g_{e'})$.

But, $s_\pi(g_e) = s_\pi(g_{e'})$, then you can repeat the last construction infinitely often in order to obtain different points x and $x' \in T$ such that $\pi(x) = \pi(x')$, which is a contradiction. ■

By using claim 3 we can define a map $\varphi' : V' \to V'$ by: for $v' \in V'$ take $\varphi'(v')$ equal to $s_{G_\pi}(e)$ where $e \in E$ is any edge with $s_\pi(e) = v'$. Since G is transitive the sets $\{\varphi'^n(v'), n \in \mathbb{N}\}$ are finite cycles, moreover this set must contain all the vertex in V'. But G is also mixing, therefore the length of this cycle must be equal to 1 and $V' = \{v\}$ proving that T' is a full shift and then (S_F, σ) is topologically conjugate to a full shift. Proving the first statement of the theorem.

Here we will not prove that λ is the maximal measure. The proof of this fact follows from the same arguments used in the proof of theorem 4.8 part (4). ■

Theorem 4.8 states that any positively expansive CA defined on $A^{\mathbb{N}}$ has the same entropy of a full shift and measure-theoretical is a full shift, however, recently M. Boyle, D. Fiebig and U. Fiebig found an example of positively expansive CA that is not topologically conjugate to a full shift. This comment shows that the problems in the one-sided and two-sided full shifts are of completely different nature.

5. Limit Sets of Cellular Automata

Among the various notions of complexity that one can associate to a cellular automaton, it is particularly interesting to study those related with its symbolic dynamics; in other words, to study the subshifts generated by the dynamics of the cellular automaton. In particular, in the previous section we have associated to each cellular automata $F : A^{\mathbb{Z}} \to A^{\mathbb{Z}}$ the canonical system (S_F, σ) (recall that it is a subshift whenever F is onto). This system take into account the complexity of F as a topological flow.

On the other hand, when F is not onto the automaton develops a nested family of subshifts,

$$..... \subseteq F^i(A^{\mathbb{Z}}) \subseteq \; ..... \subseteq F^2(A^{\mathbb{Z}}) \subseteq F(A^{\mathbb{Z}}) \subset A^{\mathbb{Z}}$$

Thus it is natural to consider the intersection

$$\Lambda(F) = \bigcap_{i=0}^{\infty} F^i(A^{\mathbb{Z}}).$$

Since it is an intersection of a decreasing family of subshifts, $\Lambda(F)$ is also a subshift that is called the *limit set of* F. Furthermore, if $S \subseteq A^{\mathbb{Z}}$ and $F(S) = S$ then $S \subseteq \Lambda(F)$. Therefore, the limit set is the maximal F-invariant subset of $A^{\mathbb{Z}}$. In particular it contains all the periodic orbits of F. A compactness argument shows that $L(\Lambda(F)) = \bigcap_{i \geq 0} L(F^i(A^{\mathbb{Z}}))$.

In this section we shall present several results concerning the complexity of F given by its limit set, or *limit complexity* of F. We mainly follow the article [33].

The study of the limit complexity of a cellular automaton was proposed by S. Wolfram [48] in 1984. First with the intention of capture the asymptotic dynamics of F and then study their attractors. In the meantime, the limit complexity has been considered following different points of view. Among others, the one given by symbolic dynamics and formal language theory, where we also consider the language $L(\Lambda(F))$. Under this perspective it is important to distinguish between two behaviours:

(1) the limit set of F is attained after a finite number of iterations of the full shift, more precisely, $\Lambda(F) = F^p(A^{\mathbb{Z}})$ for some $p \in \mathbb{N}$; in this case we say that F is a *stable* cellular automaton.

or,

(2) $\Lambda(F)$ is a truly intersection of forward images; in other words for any $p \in \mathbb{N}$, $F^p(A^{\mathbb{Z}}) \neq F^{p+1}(A^{\mathbb{Z}})$. In this case we say that the automaton F is *unstable.*

Since the subshifts $F^i(A^{\mathbb{Z}})$ for any $i \in \mathbb{N}$ are covered by the full shift $A^{\mathbb{Z}}$, which is a mixing subshift of finite type, then they must be mixing sofic systems (and therefore $L(F^i(A^{\mathbb{Z}}))$ is a regular language). It follows straightforward that the limit sets of stable cellular automata are mixing sofic systems.

Surjective cellular automata are naturally stable. The following example show one unstable cellular automata.

Example 1. Let us consider the cellular automaton F defined by its local rule $f : \{0,1\}^3 \to \{0,1\}$ such that $f(a,b,c) = 0$ if and only if $abc \neq 111$. A simple computation shows that

$$\Lambda(F) = \{x \in \{0,1\}^{\mathbb{Z}} \ / \text{ if } x_i = 1 \text{ and } x_j = 1,\ i < j, \text{ then } x(i,j) = 1^{j-i+1}\},$$

and that F is unstable.

After the article of Wolfram, different questions has been considered, some of them completely or partially solved. First of all, are the limit sets of unstable cellular automata sofic systems ? if the answer to last question is negative, which are the subshifts, or formal languages, that may be obtained as limit sets of unstable cellular automata ? is any mixing sofic system the limit set of some stable cellular automaton ? or moreover, do there exist a subshift that may be obtained as the limit set of a stable cellular automaton and of another unstable ? On the other hand, we can ask if a given cellular automaton is stable or unstable, or if its limit set is a sofic system, or a subshift of finite type, etc.

The first works considering previous questions are those of L. Hurd [24], [25]. He consider the problem whether the limit sets of unstable cellular automata are also sofic systems. In these works the author shows examples of cellular automata whose limit sets exhibit much of the same range of complexity found in language theory (particularly with respect to Chomsky's hierarchy of languages).

Theorem 5.1. ([24]) There exist cellular automata whose limit languages are,
(1) Non regular context–free languages.
(2) Non context–free, context–sensitive languages.
(3) Non recursively enumerable languages. ■

More recently, Culik, Pachl and Yu [12], and Goles, Maass and Martínez [15], have shown new classes of non recursively enumerable limit sets. Both constructions use the concept of *universal* cellular automaton. A cellular automaton F_u is *strongly universal* if it can simulate the dynamics of any other cellular automaton. In other words, given a cellular automaton F and a point x of the full shift associated to F, we can find a configuration x_u for the universal cellular automaton F_u such that it is possible to recover the dynamics $(F^i(x))_{i\in\mathbb{N}}$ from $(F_u^i(x_u))_{i\in\mathbb{N}}$. A cellular automaton is *weakly universal* if it can simulate the dynamics of any Turing machine. In both cases we say that the cellular automaton has an enormous power of computation, and then it is very complex. The main idea is to pass the complexity of the cellular automaton as a machine to the limit set. But this is not true for every universal cellular automata as is shown by the following theorem.

Theorem 5.2.
(1) ([35]) There exists a strongly universal cellular automaton whose limit language is regular.
(2) ([15]) There exists a weakly universal cellular automaton, which is not strongly universal, whose limit language is regular. ■

We have to notice that all the limit sets appearing from previous works are never mixing. Moreover, it seems that a subshift obtained as the limit set of a stable CA cannot be at the same time the limit set of an unstable CA. In this direction L. Hurd proved in [26] that subshifts of finite type cannot be obtained as the limit sets of unstable cellular automata. This assertion follows straightforward from the following proposition.

Proposition 5.3. Let $F : A^{\mathbb{Z}} \to A^{\mathbb{Z}}$ be a CA and S be its limit set. Then for any integer N there exists an integer $t(N)$ such that $F^{t(N)}(A^{\mathbb{Z}}) \subseteq S_N$ (recall that S_N is the approximation of order N of S)

Proof. Suppose the lemma is false. Then, there exists an integer N^* such that $\forall\ t \in \mathbb{N}$ we can choose $x_t \in F^t(A^{\mathbb{Z}}) \setminus S_{N^*}$ with $w_t = x_t(0, N^* - 1) \in L(F^t(A^{\mathbb{Z}})) \setminus L(S)$. Since A^{N^*} is finite, there is $w \in A^*$ such that $w_t = w$ for infinitely many $t \in \mathbb{N}$. Therefore, $w \in \bigcap_{t \in \mathbb{N}} L(F^t(A^{\mathbb{Z}})) = L(\Lambda(F)) = L(S)$, which is a contradiction. ■

As we mentioned before, if S is the limit set of a stable cellular automaton then it is a mixing sofic system. Moreover, since the image by a cellular automaton of a fixed point (for the shift) is also a fixed point, then S must contain one. Let $\tilde{a} = (a)_{i \in \mathbb{Z}}$ be a fixed point in S that is obtained as the image of another fixed point, then there exist magic words w_1 and w_2 in $L(S)$ such that $\forall\, i \in \mathbb{N}$, $w_1 a^i w_2 \in L(S)$. Such a fixed point is called a *receptive fixed point* [4]. Indeed, let $F : A^{\mathbb{Z}} \to A^{\mathbb{Z}}$ be a stable CA (of radius r) with limit set S. We can suppose $\Lambda(F) = F(A^{\mathbb{Z}}) = S$ (replacing F by a positive power). Let $\tilde{b} = (b)_{i \in \mathbb{Z}}$ and $\tilde{a} = (a)_{i \in \mathbb{Z}}$ be fixed points in $A^{\mathbb{Z}}$ and S respectively, such that $F(\tilde{b}) = \tilde{a}$. In other words $F(b^{2r+1}) = a$. Let w be a magic word of S, and $w' \in A^*$, $|w'| = |w| + 2r$, be a word such that $F(w') = w$. We deduce therefore that $F(w' b^{2r+i} w') = wua^i vw \in L(S)$ for all $i \geq 1$, where $u, v \in L(S)$. As $w_1 = wu$ and $w_2 = vw$ are also magic words of S last property is proved. When S is a SFT, the former condition is reduced to the existence of the fixed point $\tilde{a}$. We shall see below that when S is strictly sofic the receptive property of the fixed point cannot be omitted. We say that a mixing sofic system with a receptive fixed point satisfies property (H).

In what follows we shall consider two general questions,

(Q1) Which sofic systems can be the limit sets of CA (stable and unstable) ?

(Q2) If S is the limit set of some CA, what are the endomorphisms that can be obtained as the restrictions to S of a CA whose limit set is S ?

Here we give a partial answer to these questions. In the stable case the main result that we present concerns an special class of sofic systems: the class of *almost of finite type* (AFT) sofic systems. A sofic system is said to be AFT if it is the image of a transitive SFT by a factor map which is 1-to-1 on a dense open set. It was proved in [8] that a sofic system S is AFT if and only if its minimal Shannon cover $\pi : T \to S$ is left closing. Moreover, if a sofic system S is AFT then π is 1-to-1 on a dense open set. When S is AFT the set of *bad points* $M(S) = \{x \in S / |\pi^{-1}(x)| > 1\}$ and $M'(S) = \pi^{-1}(M(S))$ are proper subshifts of S and T respectively. Also, if $y_1 \neq y_2$ are points in T such that $\pi(y_1) = \pi(y_2)$, then $y_1(i) \neq y_2(i)$, $\forall\, i \in \mathbb{Z}$.

There is a graphic way to see if a sofic system S is not AFT (more precisely, to test the left closing property). We only have to look if the subgraph shown in figure 5.1 can be found in the minimal Shannon automaton of S. That is, there exist words $u, v \in L(S)$ and states $q_0, q_1, q_2 \in Q(S)$ such that $q_0 \neq q_1$, $q_0 \stackrel{u}{\to} q_0$, $q_1 \stackrel{u}{\to} q_1$, $q_0 \stackrel{v}{\to} q_2$ and $q_1 \stackrel{v}{\to} q_2$.

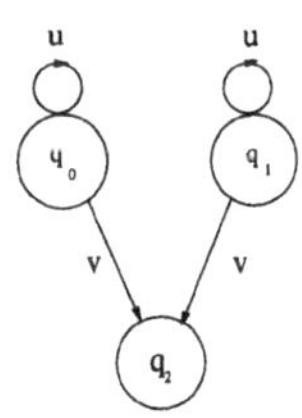

Figure 5.1

A classical example of an AFT system is the *even system* (see figure 2.1).

One of the main properties of the AFT systems is the following,

Proposition 5.4. (see [8]) Let S be an AFT system presented by the factor map $\psi : T_1 \to S$, where ψ is 1-to-1 on a dense open set and T_1 is a transitive SFT. Let $\phi : T_2 \to S$ be any factor map, where T_2 is a transitive SFT. Then, there exists a factor map $\theta : T_2 \to T_1$ such that $\phi = \psi \circ \theta$ (i.e. ϕ factors through ψ). ■

In the stable case the main result is the following.

Theorem 5.5. Any AFT system $S \subseteq A^{\mathbb{Z}}$ with property (H) is the limit set of a stable CA. ■

To prove this theorem it is necessary to state previously the same statement for subshifts of finite type.

Let $S \subseteq A^{\mathbb{Z}}$ be a SFT with property (H). If S is the limit set of a stable CA F, by definition, the restriction of the CA to S is a factor map, $G : S \to S$, and S is the direct image of the full shift by some positive power F^p of the CA, without loss of generality we can suppose $p = 1$. So, S is the limit set of a stable CA if and only if there exists a factor map $G : S \to S$ that can be extended to a factor map from $A^{\mathbb{Z}}$ to S. The factor maps G obtained in this way might have, *a priori*, some restrictions. It turns out that there are no restrictions. In fact, we prove that $id : S \to S$ (the identity map) can be extended to a factor map from $A^{\mathbb{Z}}$ to S, and therefore so can every factor map $G : S \to S$.

Theorem 5.6. Let $S \subseteq A^{\mathbb{Z}}$ be a SFT with property (H) and let $G : S \to S$ be a factor map. There exists a CA, $F : A^{\mathbb{Z}} \to A^{\mathbb{Z}}$, such that $\Lambda(F) = S = F(A^{\mathbb{Z}})$ and the restriction of F to S is G.

Proof. Suppose that S is a SFT of order 2 and that $G = id$. We denote by $\tilde{a} = (a)_{i \in \mathbb{Z}}$ the fixed point in S (given by condition (H)) and by N a transition length for S (recall that S is mixing). For each letter $b \in L_1(S)$ we can fix blocks $B(a,b), B(b,a) \in L(S)$ of length N such that $aB(a,b)b, bB(b,a)a \in L(S)$.

First, we define an auxiliar CA $\tilde{F}$. Let $x = (x_i)_{i \in \mathbb{Z}}$ be a configuration of $A^{\mathbb{Z}}$. We describe $\tilde{F}$ by its action over all the blocks $x(i,j)$ of x such that $0 < j-i+1 \leq 2N+2$, as follows:

(1) If $x(i,j) = aw_1 ....w_{2N+1}$, $aw_1 \notin L(S)$, $w_1 ......w_{2N+1} \in L(S)$, and $w_k \neq a$ $\forall\, k \in \{1, ..., 2N+1\}$, then: $\tilde{F}(x)(i, i+N+1) = aB(a, w_{N+1})w_{N+1}$.

(2) If $x(i,j) = w_1 ....w_{2N+1}a$, $w_{2N+1}a \notin L(S)$, $w_1 ......w_{2N+1} \in L(S)$, and $w_k \neq$ a
$\forall\, k \in \{1, ..., 2N+1\}$, then: $\tilde{F}(x)(j-N-1, j) = w_{N+1}B(w_{N+1}, a)a$.

(3) If $x(i,j) = aw_1 ....w_l a$, $l \leq 2N$, $aw_1 \notin L(S)$ or $w_l a \notin L(S)$, $w_1 ......w_l \in L(S)$ and $w_k \neq a$ $\forall k \in \{1, ..., l\}$, then : $\tilde{F}(x)(i,j) = a^{l+2}$.

(4) If conditions in (1),(2) and (3) do not hold:

(4.1) If $x_ix_{i+1} \in L(S)$, then $\tilde{F}(x)_{i+1} = x_{i+1}$.

(4.2) If $x_ix_{i+1} \notin L(S)$, then $\tilde{F}(x)_{i+1} = a$.

It is not difficult to see that $\tilde{F}$ is well defined, and that it is a block map of radius $2N+2$. $\tilde{F}$ has the following properties:

(1) $\tilde{F}(x)_i = a$ if $x_i = a$.

(2) $\tilde{F}(x)_i\tilde{F}(x)_{i+1} \in L(S)$ if $\tilde{F}(x)_i \neq a$ and $\tilde{F}(x)_{i+1} \neq a$.

(3) $\tilde{F}$ is the identity map on S.

(4) $\tilde{F}^2(A^{\mathbb{Z}}) = \Lambda(\tilde{F}) = S$.

Properties (1), (2) and (3) follow directly from the definition of $\tilde{F}$; let us prove property (4). By property (3), we only have to consider configurations outside S. Let $x \notin S$ and define the set $I = \{i \in \mathbb{Z} \ / \ \tilde{F}(x)_i = a\}$ (it is not empty when $x \notin S$). By property (1), $\tilde{F}^2(x)_i = a$ whenever $i \in I$, therefore, since $aa \in L_2(S)$, $\tilde{F}^2(x) \in S$ if each of the blocks $\tilde{F}^2(x)(i,j)$ such that $i < j$, $i, j \in I \cup \{-\infty, +\infty\}$, and $\{i+1, ..., j-1\} \subseteq I^c$, is in $L(S)$ (recall that S is of order two). Let $\tilde{F}^2(x)(i,j)$ be one of such blocks. By property (2), $\tilde{F}(x)(i+1, j-1) \in L(S)$. We distinguish two cases. If $j - i > 2N+1$, then, by applying rules (1),(2) or (4), as suits the case, we conclude that $\tilde{F}^2(x)(i,j) \in L(S)$. If $j - i \leq 2N+1$, then $\tilde{F}^2(x)(i,j) \in L(S)$, by rules (3) and (4). Therefore, property (4) is true and the CA $F = \tilde{F}^2$ satisfies the requirements of the theorem.

Now, consider a SFT $S \subseteq A^{\mathbb{Z}}$ of arbitrary order and suppose that $G = id$. There exists a SFT of order two, S', defined over a full shift $A'^{\mathbb{Z}}$, that is conjugate to S by the map $\rho : S \to S'$. S' also has property (H). Then, there exists a factor map $F' : A'^{\mathbb{Z}} \to S'$ extending the identity map of S'. Define the required CA as the composition $F = \rho^{-1} \circ F' \circ \bar{\rho}$, where $\bar{\rho} : A^{\mathbb{Z}} \to A'^{\mathbb{Z}}$ is an arbitrary extension of ρ. It is easy to see that F satisfies the conditions of the theorem and that its restriction to S is the identity map. By composing maps, we can conclude that theorem 5.6 is true for every factor map $G : S \to S$. ■

Let now suppose that $S \subseteq A^{\mathbb{Z}}$ is a strictly sofic system with property (H). In the last paragraph to show that a SFT with property (H) was the limit set of a stable CA we proved that any onto endomorphism of the SFT was the restricted dynamics of a stable CA over its limit set. Here it is not the case:

Proposition 5.7. [33] Let $S \subseteq A^{\mathbb{Z}}$ be a strictly sofic system with property (H). If S is the limit set of a stable CA, $F : A^{\mathbb{Z}} \to A^{\mathbb{Z}}$, then the restriction of F to S is neither right closing nor left closing. ■

The main difference with the SFT case stands in the following general observation. Let $S \subseteq A^{\mathbb{Z}}$ be a sofic system with property (H) and $G : S \to S$ be a factor map of radius r. G is also well defined over all the approximations S_N of S such that $N \geq (2r+1)$ (we say that G is *naturally defined* on S_N), then one cannot *decide locally* if a point is in S or in one of such approximations. Now, suppose that G already maps some approximation S_N, $N \geq 2r+1$, onto S, that is, $G(S_N) = S$ (it is not the case of the identity). In this case, since S_N has also property (H) (the mixing property is preserved by the approximations), then theorem 5.6 provides a CA $\tilde{F} : A^{\mathbb{Z}} \to A^{\mathbb{Z}}$ such that $\Lambda(\tilde{F}) = \tilde{F}(A^{\mathbb{Z}}) = S_N$, and the restriction of $\tilde{F}$ to S_N is the identity map. Thus, the limit set of the stable CA $F = G \circ \tilde{F}$ is S and F restricted to S is G.

In what follows we only deal with the class of *almost of finite type* sofic systems.

Let S be an AFT system, $\pi : T \to S$ be its minimal Shannon cover, and $G : S \to S$ be a factor map. By proposition 5.4, there exists a factor map $G' : T \to T$ such that $G \circ \pi = \pi \circ G'$. Since π is 1-to-1 on a dense set, G' is unique. If S is an arbitrary sofic system the lift G' should not exist, but if it exists, since π is 1-to-1 a.e., it is unique. When G' exists we call it the *canonical lift* of G.

$$\begin{array}{ccccc} & & G' & & \\ & T & \to & T & \\ \pi & \downarrow & & \downarrow & \pi \\ & S & \to & S & \\ & & G & & \end{array}$$

Now, we study the conditions that a factor map G, from an AFT system S with property (H) to itself, must satisfy in order to be *naturally defined* over some approximation SFT S_N in such a way that $G(S_N) = S$. First, we state a very general lemma.

Lemma 5.8. Let S and T be subshifts, R be a SFT and $\phi : S \to T$ be a factor map. Suppose there exist factor maps $\pi_1 : S \to R$ and $\pi_2 : R \to T$ such that $\phi = \pi_2 \circ \pi_1$, then there exists $N \in \mathbb{N}$ such that ϕ is naturally defined over the approximation SFT S_N of S and $\phi(S_N) = T$.

Proof. Without loss of generality we can suppose that R is of order two, that S is not a SFT and that the radius of ϕ, π_1 and π_2 is r. Fix $N \geq 4r + 1$ and let $x = (x_i)_{i\in\mathbb{Z}} \in S_N$. Since $N \geq 4r+1$, for all $i \in \mathbb{Z}$ the blocks $x(i-2r, i+2r) \in L(S)$, then ϕ and $\pi_2 \circ \pi_1$ are well defined over S_N (notice that $2r$ is the radius of $\pi_2 \circ \pi_1$). First, we prove that $\phi(x) = \pi_2 \circ \pi_1(x)$. For $i^* \in \mathbb{Z}$ consider a point $\bar{x} = (\bar{x}_i)_{i\in\mathbb{Z}} \in S$ such that $\bar{x}(i^*-2r, i^*+2r) = x(i^*-2r, i^*+2r)$. By hypothesis, $\phi(\bar{x})_{i^*} = \pi_2 \circ \pi_1(\bar{x})_{i^*}$, so $\phi(\bar{x})_{i^*} = \phi(\bar{x}(i^* - r, i^* + r)) = \phi(x(i^* - r, i^* + r)) = \phi(x)_{i^*}$, and $\pi_2 \circ \pi_1(\bar{x})_{i^*} = \pi_2 \circ \pi_1(\bar{x}(i^* - 2r, i^* + 2r)) = \pi_2 \circ \pi_1(x(i^* - 2r, i^* + 2r)) = \pi_2 \circ \pi_1(x)_{i^*}$. We conclude that $\pi_2 \circ \pi_1(x)_{i^*} = \phi(x)_{i^*}$, and then $\phi(x) = \pi_2 \circ \pi_1(x)$.

To finish the proof, we only have to show that $\pi_1(S_N) \subseteq R$. This follows from the sequence of equalities

$$\begin{aligned}\pi_1(x)(i^*, i^* + 1) &= \pi_1(x(i^* - r, i^* + r + 1)) = \pi_1(\bar{x}(i^* - r, i^* + r + 1)) \\ &= \pi_1(\bar{x})(i^*, i^* + 1) \in L_2(R).\end{aligned}$$

Consequently $\phi(S_N) = \pi_2(\pi_1(S_N)) = T$. ■

Lemma 5.9. Let S be an AFT system, $\pi : T \to S$ its minimal Shannon cover, and $G : S \to S$ a factor map with canonical lift G'. Then, G is naturally defined over some approximation S_N of S in such a way that $G(S_N) = S$ if and only if there exists a factor map $\theta : S \to T$ such that $G' = \theta \circ \pi$ and $G = \pi \circ \theta$.

Proof. The sufficient condition is straightforward from lemma 5.8.

Now, let $N \in \mathbb{N}$ be such that G is well defined over the SFT S_N, and $G(S_N) = S$. Since S is AFT, by proposition 5.4, there exists a factor map $\theta : S_N \to T$ such that $G = \pi \circ \theta$ over S_N (recall that S_N is also transitive). Moreover, $\theta(S) = T$. Indeed, $S = G(S) = \pi(\theta(S))$ and $\pi : \theta(S) \to S$ is a bounded-to-one factor map between two transitive sofic systems, then by lemma 2.3, $h(\theta(S)) = h(S) = h(T)$, which implies, by lemma 2.2, that $\theta(S) = T$. We conclude that $G = \pi \circ \theta$ over S, and, by the uniqueness of the lift, $G' = \theta \circ \pi$. ■

For any factor map $\psi : \bar{S} \to \bar{T}$, one defines $\mathcal{R}(\psi) = \{(x,y) \in \bar{S} \times \bar{S} \ : \psi(x) = \psi(y)\ \}$. It is not so hard to see that the condition given by lemma 5.9 is equivalent to $\mathcal{R}(\pi) \subseteq \mathcal{R}(G')$, in other words $\forall\ x \in S$, $card(G' \circ \pi^{-1}(x)) = 1$; the map θ associates to each $x \in S$ the unique point in $G' \circ \pi^{-1}(x)$, this map is continuous and shift commuting. If S is sofic (not necessarily AFT), and $G' : T \to T$ is a factor map such that $\mathcal{R}(\pi) \subseteq \mathcal{R}(G')$, then there is $\theta : S \to T$ such that G' is the canonical lift of $G = \pi \circ \theta$. In this last situation, the construction given in the first paragraph of the section and lemma 5.8 provide a stable CA whose limit set is S.

In the following lemma we exhibit an important property deduced from the existence of a receptive fixed point in the limit sets of stable cellular automata.

Lemma 5.10. Let S be an AFT system. S has a receptive fixed point if and only if its minimal Shannon cover T contains a fixed point.

Proof. To avoid confusions assume S is not of finite type.
($\Leftarrow$) Let $x = ((a,q))_{i\in\mathbb{Z}}$ be a fixed point of T, where $a \in L_1(S)$ and q is a vertex of the minimal Shannon graph of S. Let w be a magic word of S; since S is transitive and the minimal Shannon cover $\pi : T \to S$ is right resolving we can suppose that the unique vertex in the graph such that $\overset{w}{\to} q'$ is q, and that there exists $u \in L(S)$ such that $q \overset{uw}{\to} q$; uw is still a magic word, then $y = (a)_{i\in\mathbb{Z}}$ is a receptive fixed point of S (put $w_1 = w_2 = uw$ in the definition of receptive fixed points).
($\Rightarrow$) Suppose S has a receptive fixed point $y = (a)_{i\in\mathbb{Z}}$, and let w_1, w_2 be magic words associated to it. Since w_1 is magic and the minimal Shannon automaton is deterministic, there exist unique vertices q_1, q_2 such that, $\overset{w_1}{\to} q_1 \overset{a^{i_0}}{\to} q_2 \overset{a^{j_0}}{\to} q_2$, where i_0 and j_0 are minimal. To conclude we shall prove that $j_0 = 1$. Suppose $j_0 > 1$, since y is receptive, there exist two differents vertices p_1 and p_2 such that $q_2 \overset{a^{l_1}}{\to} p_1 \overset{a^{l_2}}{\to} p_2 \overset{a^{l_3}}{\to} q_2$, $l_1 + l_2 + l_3 = j_0$, $p_1 \overset{w_2}{\to}$, and $p_2 \overset{w_2}{\to}$. But since w_2 is magic both paths end in the same vertex and $\pi : T \to S$ cannot be left closing (see figure 5.1); which contradicts the assumption that S is AFT. Therefore $j_0 = 1$ and T has the fixed point $x = ((a, q_2))_{i\in\mathbb{Z}}$. ■

In order to complete the proof of the main result we shall need three lemmas from symbolic dynamics. Let $\pi_n(S)$ be the cardinality of the set of periodic points of least period n in the subshift S.

Lemma 5.11. (Embedding theorem of W. Krieger [29]) Let S be a mixing SFT and S' a subshift such that:

(i) $h(S) > h(S')$

(ii) $\pi_j(S) \geq \pi_j(S'),\ \forall\, j \in \mathbb{N}$.

Then, S' is topologically conjugate to a subshift of S.

Lemma 5.12. ([4]) Let T be a transitive SFT with positive entropy. Suppose $|\pi_J(T)| > 0$ and consider positive integers $M_1, ..., M_k$. There exists a transitive SFT S such that the following conditions hold:

(i) there is a right closing factor map $\phi : S \to T$.

(ii) S is mixing if T is mixing.

(iii) $\pi_j(S) = \pi_j(T)$ whenever $j \neq JM_i$ for any M_i and $j \neq J$.

(iv) $\pi_J(S) = \pi_J(T) - J + J\ card(\{i : M_i = 1\})$

(v) $\pi_{JM_i}(S) = \pi_{JM_i}(T) + JM_i\ card(\{j : M_j = M_i\})$ if $M_i > 1$.

Lemma 5.13. ([1]) Let S be a mixing SFT, T be a proper SFT of S and $\phi : T \to S$ be a right closing map; then ϕ can be extended to a right closing factor map $\tilde{\phi} : S \to S$.

Proof of Theorem 5.5. Assume S is not of finite type, the SFT case was already considered. Thus, since S has property (H), $h(S) > 0$. Let $\pi : T \to S$ be the minimal Shannon cover of S and M' be the set of bad points in T. Since S is AFT, M' is a proper subshift of T, and then, by lemma 2.2, we can fix an integer N such that the SFT $T^* = M'_N \subseteq T$ and $h(T^*) < h(T) = h(S)$ (recall that M'_N is the approximation of order N of M'). Let us denote $S^* = \pi(T^*)$; it is straightforward that $h(S^*) < h(S)$.

We prove there exists a factor map $G : S \to S$ such that $\mathcal{R}(\pi) \subseteq \mathcal{R}(G')$; then discussion after lemma 5.9 provides the desired CA.

First, we construct a right closing map from S^* to T. Since $h(S^*) < h(T)$, there exists an integer N_0 such that $\forall\, n \geq N_0, \pi_n(S^*) \leq \pi_n(T)$. Lemma 5.10

ensures that T has a fixed point, then we can apply lemma 5.12 to the system T with constants $J = 1$, $M_1^i = ... = M_{k(i)}^i = i$, $\forall\, i \in \{1, ..., N_0 - 1\}$, where $k(i) = \max\{\frac{\pi_i(S^*) - \pi_i(T)}{i}, 1\} + 1$. We obtain a mixing SFT T', such that $\forall\, i \geq 1$, $\pi_i(T') \geq \pi_i(S^*)$, and a right closing factor map $\psi' : T' \to T$. Finally, apply Krieger's embedding theorem to obtain a right closing map ρ from S^* to T'. Consequently, the map $\psi = \psi' \circ \rho : S^* \to T$ is right closing.

Now, define $\theta' = \psi \circ \pi : T^* \to T$, it is also right closing. Since T is mixing and T^* is a proper SFT, by means of lemma 5.13 we can extend θ' to a factor map $G' : T \to T$. It is not difficult to check that $\forall\, x \in M'$, $card(G' \circ \pi^{-1} \circ \pi(x)) = 1$, in other words $\mathcal{R}(\pi) \subseteq \mathcal{R}(G')$. To finish we have to show that G' is the lift of a factor map $G : S \to S$. But, as we remark after lemma 5.9, any factor map G' which satisfies the last condition is the canonical lift of a factor map $G : S \to S$; this completes the proof of the theorem. ■

It is important to point out that the limit sets of stable CA are not necessarily AFT. Let us consider the sofic system S, with cover T, defined by the minimal Shannon automaton in figure 5.2.

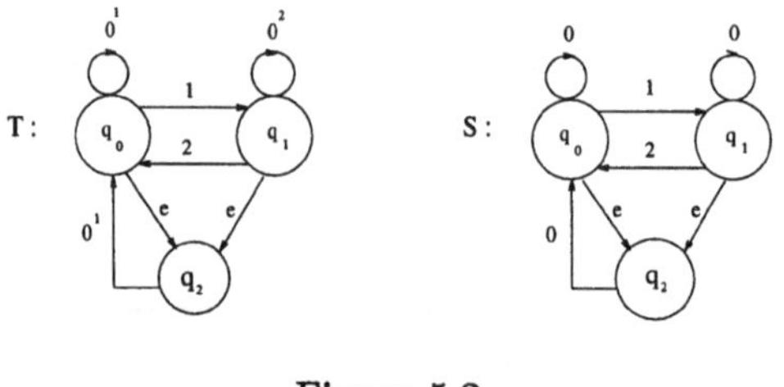

Figure 5.2

The factor $G : T \to T$ defined by its local rule: $g(ie) = e$, $\forall\, i \in \{0^1, 0^2, 1, 2\}$, $g(e0^1) = 0^1$, $g(0^1 1) = g(0^2 2) = 1$, $g(0^1 0^1) = g(0^2 0^2) = 0^1$, $g(20^1) = g(10^2) = 2$, and $g(12) = g(21) = 0^2$, induces a factor map $F : S \to S$ whose canonical lift is G. It is not difficult to show that the factor map F is the restriction of a stable CA to its limit set.

Let now turn the attention to the unstable case. We are particularly interested in the question: *Do there exist sofic systems that are limit sets of both stable and unstable CA ?*.

We do not give a general answer to this question. The main result concerns the class of *near Markov* sofic systems: a sofic system S is said to be near Markov if it is AFT and the set of bad points $M = \{x \in S / card(\pi^{-1}(x)) > 1\}$ is a finite collection of periodic points. The *even system* presented in figure 2.1 is also near Markov; its set of bad points is $M = \{^{\infty}1|_0 1^{\infty}\}$. We obtain the following result, its proof is extremely technical so we only give the proof for the even system, for the complete proof you can see [33].

Theorem 5.14. A near Markov system cannot be the limit set of any unstable CA.

Proof for the "even system": Denote $A = \{0, 1\}$ and take the sofic system S defined over A by the finite automaton in figure 2.1. Now, we will prove that S cannot be the limit set of any unstable CA. For that, we will suppose S is, and we will construct a periodic orbit for the CA living outside S.

From now, $F : A^{\mathbb{Z}} \to A^{\mathbb{Z}}$ will be a CA such that its restriction to S maps S onto S; with this assumption $S \subseteq \Lambda(F)$. r will be the neighbourhood radius of F.

Claim 1: Suppose S is the limit set of the unstable CA F, then $F(^{\infty}1|_0 1^{\infty}) = {}^{\infty}1|_0 1^{\infty}$.

Proof of claim 1: Suppose $F(^{\infty}1|_0 1^{\infty}) = {}^{\infty}0|_0 0^{\infty}$. Then, as $F(^{\infty}0|_0 0^{\infty})$ is also equal to $^{\infty}0|_0 0^{\infty}$, there exists an integer N such that $F(S_N) = S$. Thus S is the limit set of a stable CA, which is a contradiction. ■

Claim 2: Let $x_l = {}^{\infty}(0w01^l)|_0(0w01^l)^{\infty} \in A^{\mathbb{Z}}$ such that $0w \in A^*$ and $w(x_l) = 0w0 \in L(S)$. Let us suppose $F(^{\infty}1|_0 1^{\infty}) = {}^{\infty}1|_0 1^{\infty}$. Then,

(i) If $l > 5(2r+1)$ then $F(x_l) = {}^{\infty}(0\tilde{w}01^{\tilde{l}})|_t(0\tilde{w}01^{\tilde{l}})^{\infty}$ with $0\tilde{w} \in A^*$, $t \in \mathbb{Z}$, and $w(F(x_l)) = 0\tilde{w}0 \in L(S)$.

(ii) If $x_l \in S_N \setminus S$ with $N > 5(2r+1) + 2$ then $F(x_l) \notin S$.

Proof of claim 2:

(i) Let $y = {}^{\infty}1|_0 0w01^{\infty}$. As $w(x_l) \in L(S)$ and 0 is a magic word of S, then $y \in S$. Furthermore, since F is bounded to one in S, $F(y) = {}^{\infty}1|_t 0\tilde{w}01^{\infty} \in S$, for some $t \in \mathbb{Z}$. Thus, when $l > 5(2r+1)$, $F(x_l) =^{\infty} (0\tilde{w}01^{\tilde{l}})|_t(0\tilde{w}01^{\tilde{l}})^{\infty}$ where $w(F(x_l)) = 0\tilde{w}0 \in L(S)$ and $0\tilde{w} \in A^*$.

(ii) If $x_l \in S_N \setminus S$, since 0 is a magic word of S and $w(x_l) \in L(S)$, then it must happen that $01^l0 \notin L(S)$ with $l+2 > N$. Hence, by using (i), $F(x_l) =^{\infty} (0\tilde{w}01^{\tilde{l}})|_t(0\tilde{w}01^{\tilde{l}})^{\infty}$.

Now, suppose $F(x_l) \in S$, then $\tilde{l}$ must be even. Let us take $x_{l+1} \in S$ (because l is odd). Since $\tilde{l}$ is even and $l > 5(2r+1)$ $F(x_{l+1}) =^{\infty} (0\tilde{w}01^{\tilde{l}+1})|_{\tilde{t}}(0\tilde{w}01^{\tilde{l}+1})^{\infty} \notin S$. But F maps S onto S, which contradicts the last fact. ■

Let us continue with the proof of the theorem. Suppose S is the limit set of F, which is unstable, then by the first claim $F(^{\infty}1|_01^{\infty}) =^{\infty} 1|_01^{\infty}$.

Fix $N > 5(2r+1)+2$ and $t(N)$ such that $T_N = F^{t(N)}(A^{\mathbb{Z}}) \subseteq S_N$. From the second claim we deduce the existence of the configuration
$x =^{\infty} (0w01^l)|_0(0w01^l)^{\infty} \in T_N \setminus S$, with $w(x) \in L(S)$ and $0w \in A^*$.

But, again by claim 2 and the fact that $F(T_N) \subseteq T_N$, $F(x) \in T_N \setminus S$ and it is of the same type as x, then inductively we conclude that $\forall\, i \in \mathbb{N}\; F^i(x) \in T_N \setminus S$. Thus the limit set of F contains a periodic orbit outside S, which is a contradiction. ■

We want to finish this section with some questions.

First of all it would be interesting to generalize theorem 5.5 to all sofic systems with property (H). We gave some examples of non-AFT systems that are limit sets of stable CA; they were constructed from near Markov systems by adding to their minimal Shannon automata some arcs and vertices, in such a way as to prevent the resulting minimal Shannon cover from being left closing. It seems to us that this class of "pseudo-near Markov" systems is somehow archetypal, and all such systems are likely to be limit sets of stable CA.

Another interesting problem is whether the restriction of a stable CA to its limit set has a canonical lift; it is always the case for the AFT limit sets in the stable case and also for the examples of non AFT systems given in this section. Moreover, in all the examples the canonical lifts satisfy $\mathcal{R}(\pi) \subseteq \mathcal{R}(G'^p)$ for some $p \in \mathbb{N}$; this last condition seems to be the right one for constructing a stable CA whose limit set is some given sofic system having property (H).

By using arguments similar to those serving to prove Theorem 5.14, it has been proved that some non-sofic coded systems cannot be limit sets of CA [34]. Those subshifts have two properties in common, transitivity and existence of a magic word, which characterize *synchronized coded systems*. It is thus natural to ask whether the limit set of an unstable CA may be transitive (and synchronized); there is no known example by now. Also, we do not know whether there are sofic systems which are the limit sets of both stable and unstable CA.

References

[1] Ashley, J., An Extension Theorem for Closing Maps of Shifts of Finite Type, *Transactions AMS* **336**, 389-420 (1993).

[2] Adler, R., B. Marcus, Topological Entropy and Equivalence of Dynamical Systems, *Memoirs AMS* **219** (1979).

[3] Amoroso, S., Y.N. Patt, Decision Procedure for Surjectivity and Injectivity of Parallel Maps for Tessellation Structures, *J. Comput. System Sci.* **6**, 448-464 (1972).

[4] Boyle, M., Lower Entropy Factors of Sofic Systems, *Ergodic Theory and Dynamical Systems* **4**, 541-557 (1984).

[5] Blanchard, F., Cellular Automata and Transducers: A Topological View, manuscript (1993).

[6] Blanchard, F., B. Host, A. Maass, Représentation par Automate de Fonctions Continues du Tore, *Journal de Teorie des Nombres de Bordeaux*, to appear.

[7] Blanchard, F., A. Maass, Dynamical Behaviour of Coven's Aperiodic Cellular Automata, *Theoretical Computer Science*, to appear.

[8] Boyle, M., B. Kitchens, B. Marcus, A Note on Minimal Covers for Sofic Systems, *Proceedings AMS* **95**, 403-411 (1985).

[9] Blanchard, F., A. Maass, Dynamical Properties of Positively-Expansive Cellular Automata, submitted (1994).

[10] Coven, E.M., Topological Entropy of Block Maps, *Proceedings AMS* **78**, 590-594 (1980).

[11] Coven, E.M., M. Paul, Endomorphisms of Irreducible Shifts of Finite Type, *Mathematical Systems Theory* **8**, 167-175 (1974).

[12] Culik, K., J. Pachl, S. Yu, On the Limit Sets of Cellular Automata, *SIAM J. Comput.* **18**, 831-842 (1989).

[13] Denker, M., C. Grillenberger, K. Sigmund, Ergodic Theory on Compact Spaces, *Lecture Notes* **527**, Springer-Berlin (1975).

[14] Fischer, R., Sofic Systems and Graphs, *Monatshefte für Mathematik* **80**, 179-186 (1975).

[15] Goles, E., A. Maass, S. Martínez, On the Limit Set of some Universal Cellular Automata, *Theoretical Computer Science* **110**, 53-78 (1993).

[16] Gilman, R.H., Notes on Cellular Automata, preprint (1988).

[17] Gilman, R.H., Classes of Linear Automata, *Ergodic Theory and Dynamical Systems* **7**, 105-118 (1987).

[18] Glasner, E., B. Weiss, Sensitive Dependence on Initial Conditions, *Non Linearity* **6**, 1067-1075 (1993).

[19] Hedlund, G.A., Endomorphisms and Automorphisms of the Shift Dynamical System, *Mathematical Systems Theory* **3**, 320-375 (1969).

[20] Hurd, L., J. Kari, K. Culik, The Topological Entropy of Cellular Automata is Uncomputable, *Ergodic Theory and Dynamical Systems* **12**, 255-265 (1992).

[21] Hurley, M., Attractors in Cellular Automata, *Ergodic Theory and Dynamical Systems* **10**, 131-140 (1990).

[22] Hurley, M., Ergodic Aspects of Cellular Automata, *Ergodic Theory and Dynamical Systems* **10**, 671-685 (1990).

[23] Hurley, M., Attractors in Restricted Cellular Automata, *Proceedings AMS* **115**, 563-571 (1992).

[24] Hurd, L., Formal Language Characterizations of Cellular Automaton Limit Sets, *Complex Systems* **1**, 69-80 (1987).

[25] Hurd, L., The Application of Formal Language Theory to the Dynamical Behaviour of Cellular Automata, a dissertation presented to the Faculty of Princeton University in candidacy for the degree of doctor of philosophy (1988).

[26] Hurd, L., Recursive Cellular Automata Invariant Sets, *Complex Systems* **4**, 119-129 (1990).

[27] Kari, J., Rice's Theorem for the Limit Sets of Cellular Automata, *Theoretical Computer Science*, to appear.

[28] Kari, J., Decision Problems Concerning Cellular Automata, thesis of the University of Turku (1990).

[29] Krieger, W., On the Subsystems of Topological Markov Chains, *Ergodic Theory and Dynamical Systems* **2**, 195-202 (1982).

[30] Kürka, P., Languages, Equicontinuity, and Attractors in Linear Cellular Automata, preprint (1994).

[31] Lind, D.A., Application of Ergodic Theory and Sofic Systems to Cellular Automata, *Physica* **10 D**, 36-44 (1984).

[32] Lind, D.A., Entropies of Automorphisms of a Topological Markov Shift, *Proceedings AMS* **99**, 589-595 (1987).

[33] Maass, A., On Sofic Limit Sets of Cellular Automata, *Ergodic Theory and Dynamical Systems*, to appear.

[34] Maass, A., Some Coded Systems that are not Unstable Limit Sets of CA, *Cellular Automata and Cooperative Systems*, NATO-ASI series, Kluwer Ac. Publ. **396**, 433-449 (1993).

[35] Morita, K., M. Harao, Computation Universality of One Dimensional Reversible Cellular Automata, *Transactions IEICE* **72**, 758-762 (1989).

[36] Milnor, J., On the Entropy Geometry of Cellular Automata, *Complex Systems* **2**, 357-386 (1988).

[37] Nasu, M., Local Maps Inducing Surjective Global Maps of One-Dimensional Tessellation Automata, *Mathematical Systems Theory* **11**, 327-351 (1978).

[38] Nasu, M., Idecomposable Local Maps of Tessellation Automata, *Mathematical Systems Theory* **13**, 81-93 (1979).

[39] Nasu, M., An Interconnection of Local Maps Inducing onto Global Maps, *Discrete Applied Mathematics* **2**, 125-150 (1980).

[40] Nasu, M., Textile Systems for Endomorphisms and Automorphisms of the Shift, *Memoirs AMS* **546** (1995).

[41] Parry, W., Intrinsic Markov Chains, *Transactions AMS* **112**, 55-66 (1964).

[42] Parry, W., Symbolic Dynamics and Transformations of the Unit Interval, *Transactions AMS* **122**, 368-378 (1966).

[43] Shereshevsky, M.A., Ergodic Properties of Certain Surjective Cellular Automata, *Monatshefte für Mathematik* **114**, 305-316 (1992).

[44] Shereshevsky, M.A., Lyapunov Exponents for One-Dimensional Cellular Automata, *Journal of Nonlinear Science* **2**, 1-8 (1992).

[45] Shereshevsky, M.A., Expansiveness, Entropy and Polynomial Growth for Groups Acting on Subshifts by Automorphisms, *Indagationes Mathematicae* **4**, 203-210 (1993).

[46] Shirvani, M., T. Rogers, On Ergodic One-Dimensional Cellular Automata, *Communications in Mathematical Physics* **136**, 599-605 (1991).

[47] Weiss, B., Subshifts of Finite Type and Sofic Systems, *Monatshefte für Mathematik* **77**, 462-474 (1973).

[48] Wolfram, S., Computation Theory of Cellular Automata, *Communications in Mathematical Physics* **96**, 15-57 (1984).

[49] Wolfram, S., Twenty Problems in the Theory of Cellular Automata, *Physica Scripta* **9**, 170-172 (1985).

THERMODYNAMIC FORMALISM OF NEURAL COMPUTING[1]

DIMITRI PETRITIS
Institut de Recherche Mathématique
Université de Rennes I and CNRS URA 305
Campus de Beaulieu
F - 35042 Rennes Cedex
petritis@levy.univ-rennes1.fr

ABSTRACT. Neural networks are systems of interconnected processors mimicking some of the brain functions. After a rapid overview of neural computing, the thermodynamic formalism of the learning procedure is introduced. Besides its use in introducing efficient stochastic learning algorithms, it gives an insight in terms of information theory. Main emphasis is given in the information restitution process; stochastic evolution is used as the starting point for introducing statistical mechanics of associative memory. Instead of formulating problems in their most general setting, it is preferred stating precise results on specific models. In this report are mainly presented those features that are relevant when the neural net becomes very large. A survey of the most recent results is given and the main open problems are pointed out.

1. Introduction and Motivation

Comparison of the computational efficiency between the most powerful computer and the brain of the most rudimentary animal shows a really overwhelming advantage to the biological "computer" for tasks involving learning and generalisation [22]. Therefore, a considerable effort has been made to understand the brain functions and possibly mimic some of them for computational purposes. In this report,

---

[1] Work partially supported by EU network CHRX-CT93-0411.

1991 Mathematics Subject Classification: 82C32, 82B44, 82B20, 60K35

1986 Physics and Astronomy Classification Scheme: 05.20, 75.40

Key words and phrases: Neural networks, thermodynamic formalism, Gibbs measures, Hopfield model, self-averaging.

we only focus on the memory function and compare the main features between computer and brain memories. Thus,

- computer memory is localised: damage on a memory cell of the computer destroys the information contained in it; cerebral memory is diffused: even extended damage on some parts of the brain (due to trauma or tumour) does not significantly affect the contained information.
- magnetic memory is addressed: one has to know the exact location of the memory cell to recall the information contained in it; brain memory is associative: it is pointless to remember that the last friend you met on your way was the 587th person you were acquainted with; instead, it is preferable to recall her voice or the shape of her face.
- computer memory is permanent up to the moment the memory cell containing it is erased; on the contrary, there are two main types of brain memory: short term memory, erased within few minutes, and long term one, engraved up to the moment of the death, with the whole intermediate spectrum. Besides, it is very difficult to force a well memorised fact to be forgotten!
- computer memory is immediately updated in contrary to the brain memory that is progressive, an information needing repetitions in order to become permanently memorised.

Researchers from various disciplines — mathematics, computer sciences, physics, biology, psychology, linguistics, neurophysiology — are interested in the general study of brain functions. However, the epistemological backgrounds and *rationale* of these scientists are quite different. Those coming from physics or mathematics use an analytic method to describe a formidably complex system — the neural system — in terms of fairly simple, often simplist, constituting *interacting* units — the neurones. Based on their experience on phase transitions and statistical mechanics, they expect that the large scale behaviour of the system is independent of the details of the individual units so that brain functions can be expressed in terms of a small number of characteristics of the individual neurones. Those coming from medical or behavioural sciences are willing to understand how the real brain of mammals works. Finally those coming from computer sciences are interested in using the information about the brain functioning to construct more powerful computers [74].

In this survey no such issues are treated. Instead, a utilitarian point of view is adopted; the neural nets are defined as mathematical models and the consequences and implications of this definition are explained in a deductive form. Thus, mainly the mathematical aspects of the subject are treated. Although the heuristic results obtained by physicists are omnipresent and act as a *Leitmotiv* in the sequel, only rigorously established results are presented here; the reason is that there exist two excellent surveys [5, 66] on the physical results and conjectures on the subject. Moreover, not all the mathematical results are presented; a selection — dictated by personal preferences of the author — of the material is performed; the reader must always keep in mind that this text is the written version of a series of lectures taught within a finite time to an audience of real (summer) students; therefore her indulgence for omissions is implored.

Neural nets are arrays of simple processors mimicking some of characteristics of brain memory [5, 38, 39, 45, 53, 70, 95]. Several systems carry the name *neural net* nowadays; to be systematic, a system must have the following ingredients to be recognised as a neural net:

- a simple oriented graph $G = (V, E)$ where V is the set of vertices (sites), called *neurones*, and $E \subset V \times V$ is the set of edges, called *synapses*.

- a set $S \subseteq \mathbb{R}^d$ of possible *states* of every neurone. In most cases, the set S will be chosen as the binary set $S = \{-1, 1\}$ but more complicated situations may occur.

- a *configuration space*, $X = \{x : V \to S\} = S^V$, containing all the microscopic states of the system.

- a family of real variables, $J = (J_{ij})_{(ij) \in E}$, indexed by the synapses, called *synaptic efficiencies*.

- a family of real variables $= (w_i)_{i \in V}$, indexed by the neurones, called *activation thresholds*.

- a family of *post-synaptic potentials*, indexed by the neurones, defined by

$$h_i = \sum_{j \in V : (ji) \in E} x_j J_{ji} - w_i.$$

- a family of *transfer functions* $f = (f_i)_{i \in V}$, indexed by the synapses, that serves to update the value of the configuration at every neurone. This can be done either in a deterministic way — and in that case the transfer function, $f_i : \mathbb{R} \to S$, assigns a new state to the neurone i given the post-synaptic potential, h_i, that excites it by $x_i = f_i(h_i)$ —, or in a stochastic way — and in that latter case the transfer function $f_i : S \times \mathbb{R} \to [0,1]$ assigns a value to the conditional probability $\mathbb{P}(x_i = s | h_i = \eta)$ for $s \in S$ and $\eta \in \mathbb{R}$, through $\mathbb{P}(x_i = s | h_i = \eta) = f_i(s, \eta)$.

Definition 1.1. We call *neural net* the system (G, X, J, w, h, f), *i.e.* an oriented graph G, a configuration space X, two families of real variables — the synaptic efficiencies J and the activation thresholds w — a family of post-synaptic potentials h and a family of transfer functions f. According to the mode of transfer, the net is termed deterministic or stochastic.

Remark. The graph G being simple and oriented, there exists a natural order induced by the orientation. For finite graphs G, we can stratify the set of vertices in the following manner:

$$V_0 = \{i \in V : \forall j \in V, (ji) \notin E\};$$

this set represents the *sensor neurones* that receive the *external stimuli*. The intermediate strata are defined recursively, by

$$V_k = \{i \in V \setminus (V_0 \cup V_1 \cup \cdots \cup V_{k-1}) : \exists j \in V_{k-1} \text{ such that } (ji) \in E\}.$$

The last layer, $L = \sup\{k : V \setminus \cup_{i=0}^{k} V_i) \neq \emptyset\} + 1$, gives rise to the stratum V_L of *motor neurones* that communicate to the external world the result of the computation. This stratification of the vertex set induces a natural stratification of the configuration space $X = \oplus_{l=0}^{L} X_l$, with $X_l = \{x : V_l \to S\} = S^{V_l}$.

The configuration over the set V_0 must be fixed *ad hoc* but the configuration over all the other strata is determined by the neural network updating rules. This updating can be done either in a *synchronous* way, when whole batches of neurones change simultaneously their internal configuration according to the post synaptic potential they receive from their "parent" neurones, or in an *asynchronous* way, when to each neurone is attached an internal clock — independent of the clock of the other neurones — that commands the moment of updating. In the case of a synchronous net, the time evolution can be studied as a discrete time dynamical system; for asynchronous nets, continuous time evolution is more appropriate.

Given a configuration $x \in X_0$ at the sensor layer, the net returns a configuration $y \in X_L$ at the motor layer, implementing thus a mapping $F : X_0 \to X_L$ that is completely determined by $x \mapsto y = F(x)$ in terms of the set of parameters $J = (J_{ij})_{(ij)\in E}$ and $w = (w_i)_{i\in V}$. It will be convenient in the sequel to consider the set of all possible parameters J and w as a space Θ, generic points, $\theta = (J, w) \in \Theta$, of which are meant to represent a given realisation of the network. This space is termed *control space* and it will be eventually equipped with a probability measure. The choice of the θ, fixing the parameters of the network, controls (defines) completely the map F. To stress this control, we write, when necessary, F_θ for this map.

At this level of generality, it is difficult to implement the network, to understand its functioning, and to decide whether it is advantageous to use a neural computer versus a conventional one. We must therefore specify the network more precisely: this will be done by studying particular examples.

2. Examples of Neural Networks

A whole spectrum of neural networks is introduced; they are classified according to the nature of their state space, their architecture, and their transfer function.

Since implementation on a digital computer proceeds always by discretisation, very often, it is enough to consider binary networks, *i.e.* $S = \{-1, 1\}$. These networks have a configuration space reminiscent of the Ising configuration space in statistical mechanics. However, some more general single-neurone internal states spaces are used. For instance, q-states Potts neural nets [33] have $S=\{0,\cdots,q-1\}$, XY-neural nets [89] have $S=\mathbb{T}^1$, and so on.

The second characteristic serving to classify the nets is their architecture. This is defined mainly by the edge structure of the underlying graph. However, architecture can also be a dynamical characteristic. As a matter of fact, edges serve to index the synaptic efficiencies. Now, if for a given graph $G=(V,E)$, the synaptic efficiencies $J_{ij}\not\equiv 0$ only for all $\{ij\}\in E'\subset E$, it is the set E' that defines the architecture and not the set E. In the same spirit, the architecture can be even a random characteristic as it may happen in the case of randomly diluted nets [15], where each synaptic efficiency is multiplied by a random variable that can take values 0 or 1 independently on every edge of the graph, leading thus to a bond percolation cluster sub-graph of E.

The nature of the transfer function must also be taken into account. In view of practical applications, it is convenient to be able to parallelise the computations; therefore it is natural to consider the same transfer function function all over the network. For deterministic systems, we can choose the non-linear function $f:\mathbb{R}\to S$ defined by

$$f(s)=\operatorname{sgn}(s)=\begin{cases}1 & \text{if } s\geq 0\\ -1 & \text{if } s<0.\end{cases}$$

In several applications some continuity properties are required; in that case instead of the sharp sgn function a smoothed sigmoidal version can be used, for instance $f(s)=\tanh(\beta s)$, for some real parameter β.

Finally, the synchronisation must be specified to decide whether the updating follows a synchronous or an asynchronous updating schedule. In case of synchronous updating and for some architectures, nets may have different evolution depending on their parallel or sequential updating [63, 79, 80, 103].

Having these characteristics in mind we present some of the most commonly studied neural nets.

2.1. THE NEURAL NET OF MCCULLOCH AND PITTS

This model [65], introduced in 1943, is composed by a semi-infinite repetition of N neurones, *i.e.* its vertex set is $V=\{1,\ldots,N\}\times\mathbb{N}$ and its space state $S=\{-1,1\}$

is binary. The particular structure of the underlying graph allows to represent each vertex $v \in V$ as $v = (i, t)$ with $i \in \{1, \ldots, N\}$ and $t \in \mathbb{N}$. The latter index is interpreted as time. If $v = (i, t)$ and $v' = (j, t+1)$, the corresponding synaptic efficiency $J_{v,v'}$ is denoted $J_{ij}^{(t+1)}$ and the post-synaptic potential, $h_{v'} \equiv h_j(t+1)$, is expressed in terms of the configurations by

$$h_j(t+1) = \sum_{i=1}^{N} x_i(t) J_{ij}^{(t+1)} - w_j^{(t+1)}.$$

The computation is then expressed in terms of the discrete time dynamics

$$x_j(t+1) = \text{sgn}(h_j(t+1)).$$

Sequential updating is used to induce a dynamical evolution $X^{(t)} \to X^{(t+1)}$ where $X^{(t)} = S^{V_t}$ and V_t is the set of neurones involved at time t. The stratum S^{V_t} of the configuration space being isomorphic to S^N for every t, one time step of the network is a mapping $T : S^N \to S^N$ and the evolution can be regarded as a trivial (*i.e.* deterministic) Markov chain on S^N defined for $a, b \in S^N$ by

$$p_{ab} = \mathbb{P}(Y_{n+1} = b | Y_n = a) = \begin{cases} 1 & \text{if } b = Ta \\ 0 & \text{otherwise.} \end{cases}$$

The reason for introducing such a trivial probabilistic object is that it can be easily generalised to tackle stochastic dynamics introduced in subsequent sections. When $N \to \infty$, the previous evolution is a discrete dynamical system $[0,1] \to [0,1]$. We are interested in the asymptotic behaviour of T^t when $t \to \infty$; (for the mathematical treatment of such evolutions the reader may consult [23] for instance). It has been shown that the McCulloch and Pitts network is computationally equivalent to a Turing machine. It is not however evident that it offers a more efficient alternative to the computation of general functions than a universal computer.

2.2. THE SIMPLE PERCEPTRON

This is the most elementary neural net, introduced [86] in 1962. It is a binary, *i.e.* $S = \{0, 1\}$, synchronous net over a finite bipartite graph whose vertex set is $V_0 \oplus V_1$ with $V_0 = \{1, \cdots, N\}$ and $V_1 = \{1\}$, *i.e.* there are N sensor neurones and one motor neurone. The edge set is $E = \{(i,j) : i \in V_0, j \in V_1\}$ and the transfer function is plainly $f = \frac{1+\text{sgn}}{2}$. Therefore, the simple perceptron implements a Boolean function of N inputs and one output.

A fundamental question, for using neural computing, is whether all Boolean functions can be implemented by a single perceptron *i.e.* whether, for every Boolean function with N entries, $G_k : S^N \rightarrow S$, with $k = 1, \cdots, 2^{2^N}$, is it possible to chose the control parameters $\theta^{(k)}$ (synaptic efficiencies and activation thresholds) of simple perceptron so that

$$G_k \equiv F_{\theta^{(k)}}, \quad \forall k = 1, \cdots, 2^{2^N}.$$

The answer to this question is negative as it can be shown for the exclusive or (XOR) Boolean function of two entries. In fact, for the simple perceptron, the input configuration space X_0 has a vector space structure and this endows the control space $\Theta \equiv X_0'$ with the dual linear structure. In other words, a choice θ of control parameters defines a linear functional $\theta \in \Theta \equiv X_0'$ and the post-synaptic potential is the action of this functional to the entry configuration[2]

$$h = \sum_{i=1}^{N} J_i x_i = \langle \theta, x \rangle.$$

Definition 2.1. Let A and B be two subsets of the space X_0. We say they are *linearly separable* if there exists a linear functional $\theta \in X_0'$ such that

$$\forall x \in A, \forall y \in B, \quad \langle \theta, x \rangle < \langle \theta, y \rangle.$$

If no such linear functional exists, the sets are called non-separable.

[2] The activation threshold can be incorporated in this writing by adding a $N+1$ input fixed to the value 1.

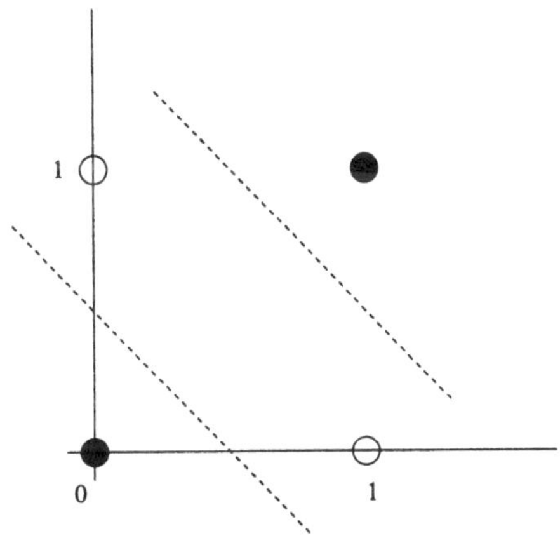

Figure 1. The figure gives the truth table of the XOR function. Full blob (or number 0) stands for `false` and empty blob (or number 1) for `true`. It is evident that there is no single straight line splitting the plane into two regions containing single colour points; two such lines are necessary. The truth space of the XOR function is not linearly separable.

It is immediate to see (figure 1) that the truth table of the Boolean function XOR with two entries splits the space X_0 into two regions needing two linear functionals to be separated instead of just one. As a consequence, this function cannot be implemented by a single perceptron. This remark was at the origin of the oblivion into which felt the neural computing for more than two decades.

2.3. THE MULTILAYERED PERCEPTRON

The multilayered perceptron [88] is a binary neural net with sharp transfer function. It is composed by a finite simple oriented graph of processors arranged in various layers $l = 0, \dots, L$; each layer is composed by N_l neurones so that V_l is isomorphic to $\{1, \dots, N_l\}$ and the configuration space, X, is stratified, $X = \otimes_{l=0}^{L} X_l$, where

$$X_l = \{x : V_l \cong \{1, \dots, N_l\} \to S\} = S^{N_l}.$$

Only edges connecting processors of a given layer, l, with processors of the next layer, $l+1$, appear and we denote by $J_{ij}^{(l+1)}$ the synaptic efficiency between the site i of the l-th later with the j site of the $l+1$-th layer. Thus the post-synaptic potential reads

$$h_j^{(l+1)} = \sum_{i \in V_l} x_i J_{ij}^{(l+1)} - w_j^{(l+1)}$$

and the deterministic dynamics

$$x_j^{(l+1)} = \mathrm{sgn}(h_j^{(l+1)}), \quad \text{for} \quad j \in V_{l+1}.$$

The interest of the multilayered perceptron stems on the fact that it can implement all Boolean function with an arbitrary number of entries. More precisely, it is shown (see [70] for instance) the following

Theorem 2.2. Let $F : \{-1,1\}^N \to \{-1,1\}$ be a Boolean function with N entries and one output. Then, there exists a two layered network with binary neurones, with a layer of N sensor neurones, a hidden layer of 2^N neurones, and a layer of one motor neurone that implements the function F without error.

Remark. What remains an open question however is whether there is an optimal size for the intermediate layer not saturating the bound required by the previous theorem. For specific Boolean functions the answer is affirmative; for instance, the function XOR with two entries can be implemented on a two layered network with only two intermediate neurones instead of the four required by the existence theorem. It is not clear whether this bound is really non saturated for a general Boolean function with an arbitrary number of inputs.

2.4. FULLY CONNECTED COMMITTEE MACHINE

This network was introduced in [72]. It is a special case of a two-layered perceptron with binary neurones and sharp transfer functions. The input layer has N sensor neurones, the intermediate layer has K hidden neurones and the output layer has one motor neurone. There is a first class edge connecting every input neurone to every hidden neurone and a second class edge connecting every hidden neurone to the output neurone. The first class edges carry synaptic efficiencies J_{ij} with

$i = 1, \cdots, N$ and $j = 1, \cdots, K$ so that

$$x_j = \text{sgn}\left(\sum_{i=1}^{N} x_i J_{ij}\right), \quad i = 1, \cdots, N \quad j = 1, \cdots, K$$

and the second class of edges carry synaptic efficiencies that are all equal to one so that

$$x = \text{sgn}\left(\sum_{j=1}^{K} \text{sgn}(\sum_{i=1}^{N} x_i J_{ij})\right).$$

Thus, a fully connected committee machine [1, 91] is a fully connected simple perceptron of N entries and K outputs feeding a K entries majority rule voting machine[3].

2.5. XY-NETWORKS

This is a continuous state space network with $S = \mathbb{T}^1$ introduced in [55]. The vertex set is $V = \{1, \cdots, N\} \times \mathbb{N}$. Thus, as it is the case for the McCulloch and Pitts network, sites have a spatial and a temporal component, namely $v = (i, t)$, with $i \in \{1, \cdots, N\}$ and $t \in \mathbb{N}$. Edges connect sites of a given layer with all sites of the next layer. Each neurone is attached an activation threshold $w_i(t)$ and each edge a synaptic efficiency $J_{ij}^{(t+1)}$. The novel feature of this network is the way the updating is performed:

$$x_i(t+1) = \sum_{j=1}^{N} J_{ij}^{(t+1)} \sin(x_i(t) - x_j(t)) - w_i^{(t+1)} \mod 2\pi.$$

Usually, the activation threshold of this model is a Gaussian random variable independent from site to site.

This network has also a continuous time version [89] for asynchronous updating, leading in an underlying space-time structure of the form $V = \{1, \cdots, N\} \times \mathbb{R}^+$. In this case the updating is through the stochastic differential equation

$$\frac{dx_i(t)}{dt} = \sum_{j=1}^{N} J_{ij}(t) \sin(x_i(t) - x_j(t)) - w_i(t) \mod 2\pi.$$

[3] To avoid any indeterminacy, we can chose K to be odd.

Notice however that in this case the model resembles more to a field-theoretical model and less to a statistical mechanical one. In the usual case of random activation threshold, $w_i(t)$ is a white noise generalised random process, independent for different i's.

2.6. SHORT RANGE FINITE-DIMENSIONAL NETWORK

For layered networks, it is convenient to interpret a given layer as the spatial extent of the network and the passage from a layer to the next one as a time evolution. The common feature of all the nets introduced so far is that all neurones of a given layer intervene to define the state of a single neurone of the next layer; in this sense the previous models are long-range — as a matter of fact mean-field models (see [82] for instance for definitions) — in the statistical mechanics terminology.

Another possibility should be to define a layered network $V = \oplus_{l \in \mathbb{N}} V_l$ where each layer V_l is isomorphic to a finite subset of a regular D-dimensional lattice,

$$V_l \cong [-N, N]^D \cap \mathbb{Z}^D.$$

The novel feature is the "short range" edge structure. In this model, a pair of vertices (v, v'), where $v = (i, t)$ and $v' = (i', t'))$, with $i, i' \in [-N, N]^D \cap \mathbb{Z}^D$ and $t, t' \in \mathbb{N}$, belongs to the edge set E if, and only if, $t' = t + 1$ and $|i - i'| = 1$, where $|\cdot|$ denotes the Euclidean distance in $\mathbb{Z}^D$. The simplest network has binary neurones but more complicated nets can be defined. The edge set indexes a family of synaptic efficiencies and the (synchronous) update is performed according to the usual formula. Such short range models are studied in [73] or [19].

2.7. RANDOMLY DILUTED NETWORKS

These networks can be defined as a modification of any of the previously defined nets. Edges of the underlying graph index not only a family of synaptic efficiencies but also a family of independent identically distributed random variables $(c_e)_{e \in E}$, taking values in $\{0, 1\}$. The effective synaptic efficiency is $J_e c_e$. Therefore, the edge set that really contributes to the network architecture is the set $E_1 = \{e \in E : c_e = 1\} \subset E$. Since this procedure is equivalent in erasing some edges, the resulting network is termed randomly diluted.

3. Learning Algorithms

We have seen that a neural net on a finite simple graph is the implementation of a mapping $F : X_0 \to X_L$ depending on the particular realisation $\theta \in \Theta$ of the control parameters. To stress out this dependence, we shall use in the sequel the notation F_θ for the mapping implemented by the realisation θ.

In this section, we address the inverse problem; namely the problem of choosing the realisation of the control parameters so that the network implements a given mapping, provided the implementation is possible. The direct problem will be discussed in the following sections.

3.1. SUPERVISED LEARNING

The most convenient description is the statistical one. The space Θ contains all possible realisations of the network. In the absence of any additional information, the parameters can have arbitrary values. In general, we can assume that in such a situation there is an *a priori* measure μ_0 on Θ, that is reasonable to choose non-atomic and having support on the whole space Θ.

Supervised learning will be interpreted as a modification of this measure μ_0 in such a way that it will become concentrated on smaller and smaller sets [101]. To be more specific, we stick to the deterministic multilayered perceptron with binary neurones and constant transfer function $f = \text{sgn}$. For a given realisation θ, we denote, as usual, by F_θ the mapping implemented by the network. We identify in the sequel any mapping $g : X_0 \to X_L$ with its graph

$$g \equiv \{\Xi = (x, y) \in X_0 \times X_L : \Xi = (x, g(x))\}.$$

Definition 3.1. A *training set* for the mapping $g : X_0 \to X_L$ is a finite subset, L_g, of the graph of g, *i.e.*

$$L_g = \{\Xi^\alpha = (x^\alpha, y^\alpha) : y^\alpha = g(x^\alpha)\}_{\alpha=1,\ldots,A}.$$

We shall say that the network has been totally trained by the training set L_g if the control parameters have been adjusted in such a manner that[4] $F_\theta |_{L_g} = g |_{L_g}$.

Such a total training is however very demanding in terms of time; moreover, it reduces the generalisation capabilities of the network. It is much more efficient to allow for a small number of errors. To be more specific, denote by d a natural distance on the configuration space stratum X_L, for instance the Hamming distance, defined for every two configurations x and y as the number of sites where they differ

$$d(x,y) = \sum_{i=1}^{N_L} \frac{(x_i - y_i)^2}{4},$$

and define a map $\mathcal{H}_A : \Theta \to \mathbb{R}^+$ by

$$\mathcal{H}_A(\theta) = \sum_{\alpha=1}^{A} d(y^\alpha, F_\theta(x^\alpha)).$$

The function is called *total learning error* and attains its minimal value, 0, when the net is totally trained on L_g. Otherwise it takes a positive value that counts the number of errors between the graphs of g and of F_θ for the specific realisation of the control parameters. Any sensible learning algorithm can be viewed as an algorithm searching for the minimum of $\mathcal{H}_A$.

3.2 DETERMINISTIC LEARNING ALGORITHMS

Deterministic algorithms are totally specified sequences of control parameters $(\theta_n)_{n\in\mathbb{N}}$ exploring a subset of Θ. It is expected that such sequences converge towards the global minimum of $\mathcal{H}_A$, *i.e.* that

$$\lim_{n\to\infty} \theta_n = \overline{\theta} = \arg\min_{\theta\in\Theta} \mathcal{H}_A(\theta).$$

However, for such a convergence to hold, some additional conditions are needed. Typically, convexity and differentiability of $\mathcal{H}_A$ are required. Differentiability can

[4] A slightly abusive notation is used here; restriction of a function on L_g actually meaning restriction to the set of the first co-ordinates of points composing L_g.

be replaced by continuity and subdifferentiability; on the contrary, convexity is a very important condition.

3.2.1. *Gradient and subgradient algorithms and error back-propagation.* Let $\mathcal{H} : \Theta \rightarrow \mathbb{R}^+$ be a numerical function, defined on a finite-dimensional vector space Θ, which we want to minimise. Suppose moreover that $\mathcal{H}$ is differentiable. Obviously, if $\bar{\theta} = \arg\min_{\theta \in \Theta} \mathcal{H}_A(\theta)$, then $\mathcal{H}'(\bar{\theta}) = 0$. The *gradient algorithm* reads [8]

Algorithm 3.2 (Gradient minimisation algorithm)

FIX some numerical sequence $(\delta_n)_{n \in \mathbb{N}}$ of positive numbers.
INITIALISE $n \leftarrow 0$
CHOOSE some arbitrary $\theta_0 \in \Theta$.
REPEAT UNTIL $\mathcal{H}'(\theta_n) = 0$
{ $\qquad \theta_n \leftarrow \theta_n - \delta_n \frac{\mathcal{H}'(\theta_n)}{\|\mathcal{H}'(\theta_n)\|}$
$\qquad n \leftarrow n + 1.$
}

This algorithm converges in general to local minima. We have however the following

Theorem 3.3. Let the sequence $(\delta_n)_{n \in \mathbb{N}}$ of the previous algorithm be chosen so that

1. $\lim_{n \rightarrow \infty} \delta_n = 0,$
2. $\sum_{n \in \mathbb{N}} \delta_n = +\infty,$

and assume that $\mathcal{H} : \Theta \rightarrow \mathbb{R}$ is a convex, differentiable, bounded from below function. Then the sequence

$$y_k = \min_{n=0,\ldots,k} \mathcal{H}(\theta_n),$$

where (θ_n) is the sequence defined in the previous algorithm, converges to the infimum of $\mathcal{H}$, namely,

$$\lim_{k \rightarrow \infty} y_k = \inf_{\theta \in \Theta} \mathcal{H}(\theta).$$

Now, the differentiability of the function $\mathcal{H}$ is a very stringent condition; very often it can be verified for a distance used to define $\mathcal{H}$ and can fail for a slight modification of the distance function. It is therefore wishable to get rid of this condition and to replace it by less restrictive and more stable conditions of continuity and subdifferentiablity. Recall that gradient, when it exits, is a linear form on the tangent space. According to [85], define then

Definition 3.4. Let $\mathcal{H} : \Theta \to \mathbb{R}$ be a convex function. A vector θ^* is called a *subgradient* of $\mathcal{H}$ at θ if

$$\mathcal{H}(\theta') \geq \mathcal{H}(\theta) + \langle \theta^*, \theta' - \theta \rangle, \quad \forall \theta' \in \Theta.$$

The set of all sugradients of $\mathcal{H}$ at θ is called the *subdifferential* at θ and is denoted $\partial\mathcal{H}(\theta)$.

For continuous functions, the minimisation algorithm becomes

Algorithm 3.5 (Subgradient minimisation algorithm)

FIX some numerical sequence $(\delta_n)_{n\in\mathbb{N}}$ of positive numbers.
INITIALISE $n \leftarrow 0$
CHOOSE some arbitrary $\theta_0 \in \mathrm{Dom}(\mathcal{H})$ and some subgradient $p_0 \in \partial\mathcal{H}(\theta_0)$.
REPEAT UNTIL $p_n = 0$

```
{       CHOOSE some p_n ∈ ∂H(θ_n)
        θ_n ← θ_n − δ_n p_n/||p_n||
        n ← n + 1.
        }
```

The convergence of this algorithm is guaranteed by the following

Theorem 3.6. Let the sequence $(\delta_n)_{n\in\mathbb{N}}$ of the previous algorithm be chosen so that

1. $\lim_{n\to\infty} \delta_n = 0$, and
2. $\sum_{n\in\mathbb{N}} \delta_n = +\infty$.

Assume that $\mathcal{H} : \Theta \to \mathbb{R} \cup \{+\infty\}$ is a convex, bounded from below function and that the interior of $\mathrm{Dom}(\mathcal{H})$ is non empty. Then the sequence

$$y_k = \min_{n=0,\ldots,k} \mathcal{H}(\theta_n),$$

where (θ_n) is the sequence defined in the previous algorithm, converges to the infimum of $\mathcal{H}$, namely,

$$\lim_{k\to\infty} y_k = \inf_{\theta\in\Theta} \mathcal{H}(\theta).$$

Proof. See [8] for instance. ∎

Application of these minimisation algorithms gives rise to the so called *error back propagation adaptation scheme.* To illustrate the method, consider a $L+1$-layered perceptron with N neurones at every layer and a smooth transfer function f. Recall that we want to minimise

$$\begin{aligned}\mathcal{H}_A(\theta) &= \sum_{\alpha=1}^{A} d\left(y^\alpha, F_\theta(x^\alpha)\right) \\ &= \frac{1}{4}\sum_{\alpha=1}^{A}\sum_{i=1}^{N} \left(y_i^\alpha - F_\theta(x^\alpha)_i\right)^2 .\end{aligned}$$

Assume to simplify notation that the activation thresholds are identically vanishing all over the network. Starting from a given initial realisation of the control parameters,

$$\theta = \left(J_{ij}^{(l)}\right)_{i,j=1,\ldots,N}^{l=0,\ldots,L},$$

we wish to follow the gradient algorithm, that is to modify synaptic efficiences according to the formula

$$J_{ij}^{(l)} \leftarrow J_{ij}^{(l)} - \delta_1 \frac{H_{ij}^{(l)}}{|H_{ij}^{(l)}|}$$

where

$$H_{ij}^{(l)} = \frac{\partial \mathcal{H}_A(\theta)}{\partial J_{ij}^{(l)}}.$$

The practical problem to solve therefore is to compute the partial derivative in $H_{ij}^{(l)}$. To update the control parameters, it proves convenient to start from the last layer L and continue backwards to the 0-th layer, hence the name error back propagation, namely

$$\begin{aligned} H_{ij}^{(L)} &= \frac{\partial \mathcal{H}_A(\theta)}{\partial J_{ij}^{(L)}} \\ &= \frac{1}{4} \sum_{\alpha=1}^{A} \frac{\partial}{\partial J_{ij}^{(L)}} \sum_{k=1}^{N} (y_k^\alpha - F_\theta(x^\alpha)_k)^2 \\ &= -\frac{1}{2} \sum_{\alpha=1}^{A} \left(y_j^\alpha - f(h_j^{\alpha,(L)})\right) f'\left(h_j^{\alpha,(L)}\right) x_i^{\alpha,(L-1)}, \end{aligned}$$

where

$$h_j^{\alpha,(L)} = \sum_{k=1}^{N} x_k^{\alpha,(L-1)} J_{kj}^{(L)}$$

is the post-synaptic potential due to the 0-th layer configuration x^α. After the modification on $J_{ij}^{(L)}$ is completed for the L-th layer, the procedure is restarted with the layer $L-1$ and then again up to the first layer. It is thus established that gradient descent search leads to the error back propagation learning algorithm [45, 53].

3.2.2. *Newton algorithm and progressive learning.* First recall the principle of Newton's method for the search of the zero of a smooth function $\mathcal{H} : \Theta \to \mathbb{R}$, under the assumption that $\mathcal{H}'(\theta)$ is invertible. The methods reads

Algorithm 3.7 (Newton's one-dimensional method)

INITIALISE $n \leftarrow 0$

CHOOSE $\delta > 0$ and $\theta_0 \in \Theta$.

REPEAT until $\mathcal{H}(\theta_n) = 0$

{ $\theta_n \leftarrow \theta_n - \delta \frac{\mathcal{H}(\theta_n)}{\mathcal{H}'(\theta_n)}$

$n \leftarrow n + 1.$

}

Proof. It is an elementary exercise (see exercise 6, page 380 of [57] for instance) to show that under C^2 and boundedness conditions, the above sequence converges to a zero, *i.e.*

$$\lim_{n\to\infty} \theta_n = \overline{\theta}$$

with

$$\mathcal{H}(\overline{\theta}) = 0. \quad \blacksquare$$

The method can be easily generalised to the multi-dimensional case with only notational complications (the derivative is now a linear operator on the tangent space).

Turn now to the specific problem we have to solve, namely adjust the control parameters θ so that the net becomes totally trained on some training set. In other words, we have to find a realisation $\overline{\theta}$ such that $\mathcal{H}_A(\overline{\theta}) = 0$. Now,

$$\begin{aligned}\mathcal{H}_A(\theta) &= \sum_{\alpha=1}^{A} d\left(y^\alpha, F_\theta(x^\alpha)\right) \\ &= \frac{1}{4}\sum_{\alpha=1}^{A}\sum_{i=1}^{N} \left(y_i^\alpha - F_\theta(x^\alpha)_i\right)^2,\end{aligned}$$

and since every term in the above sum is positive, it corresponds to an additional constraint on the control parameters. Starting from an arbitrary realisation $\theta_0 = (J_{ij}^l)$, change the parameters by $\Delta^1 J_{ij}^l$, where

$$\Delta^1 J_{ij}^l = \frac{\delta}{4}\left[\frac{\partial}{\partial J_{ij}^l}\sum_m \left(y_m^1 - F_\theta(x^1)_m\right)^2\right]^{-1} \sum_{i=1}^{N} \left(y_i^1 - F_\theta(x^1)_i\right)^2,$$

for $i, j = 1, \ldots, N$ and $l = 0, \ldots, L$. After this change, the parameters are "closer" to those corresponding to a net having learnt the first example. Now, repeat the same procedure with the other examples

$$\Delta^\alpha J_{ij}^l = \frac{\delta}{4}\left[\frac{\partial}{\partial J_{ij}^l}\sum_m \left(y_m^\alpha - F_\theta(x^\alpha)_m\right)^2\right]^{-1} \sum_{i=1}^{N} \left(y_i^\alpha - F_\theta(x^\alpha)_i\right)^2,$$

for $i, j = 1, \ldots, N$ and $l = 0, \ldots, L$ and for $\alpha = 2, \ldots, A$. When all the examples are scanned consider the realisation

$$\theta_1 = \left(J_{ij}^l + \sum_{\alpha=1}^{A} \Delta^\alpha J_{ij}^l \right)_{ij}^l$$

as the new starting point and iterate. This algorithm corresponds to a *progressive learning* procedure.

3.3. A DIGRESSION: NEED FOR NEW ALGORITHMS

We have seen that the two previous deterministic algorithms converge under rather stringent conditions, convexity and continuity for the subgradient algorithm, derivability and inversibility of the derivative operator for the Newton's algorithm. For small nets we can hope or even explicitly check that such conditions hold. For large nets however, it is unreasonable even to think that such conditions can be valid. Experience from the area of spin glasses and numerical evidence show that they almost surely fail.

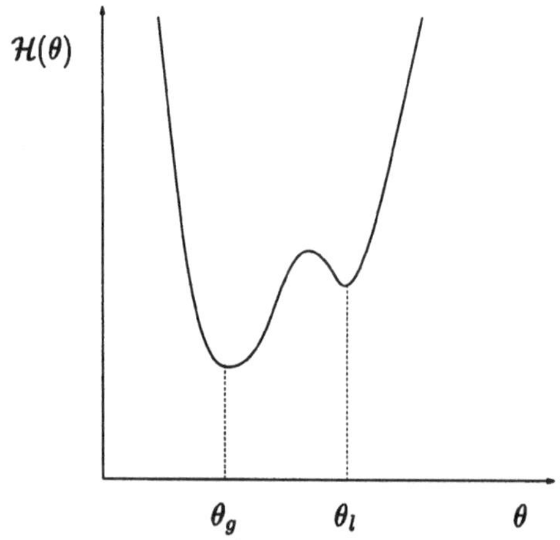

Figure 2. When the function to minimise is not convex, depending on the initial point θ_0, the algorithm will eventually converge either to the local minimum θ_l or the global one θ_g.

To be more specific, consider the trivial counter-example, depicted in figure 2, of a smooth function, bounded from below, but not convex. It is immediate to see that depending on the starting point for the sequence

$$\theta_{n+1} = \theta_n - \delta_n \frac{\mathcal{H}'(\theta_n)}{\|\mathcal{H}'(\theta_n)\|}$$

the algorithm converges to a global or a local minimum. Intuitively, the sequence θ_n can be thought as the positions of a ball without kinetic energy falling with friction into the potential well defined by the function $\mathcal{H}$. The ball gets trapped to the first local minimum encountered. As a matter of fact, this trapping phenomenon is a common feature of all deterministic algorithms. For realistic functions $\mathcal{H}$ we need therefore some new algorithms, improved by stochastic considerations, like *stochastic gradient* or *simulated annealing.*

3.3.1. *Stochastic gradient algorithms.* For simple cases as the one exemplified above, an alternative is provided by the stochastic gradient algorithm: it is a stochastic process $(X_n)_{n\geq 0}$ with values in the space Θ, defined by the sequence

$$X_{n+1} = X_n - \delta_{n+1}\mathcal{H}'(X_n) + \epsilon_{n+1}\xi_{n+1}$$

where $(\xi_n)_{n\geq 1}$ is a sequence of independent identically distributed variables and (δ_n) and (ϵ_n) are two deterministic (or previsible with respect to the natural filtration $\sigma(\xi_1, \ldots, \xi_n)$) sequences tending slowly to zero. When the speed of convergence to zero of these sequences is properly chosen, the stochastic process (X_n) converges to a random variable X_∞, distributed according to a law charging with large probability small intervals containing in their interior the absolute minimum of $\mathcal{H}$. We don't wish to give further details on the convergence criteria since even this algorithm has a limited field of applications in small nets. It is only useful to remark that the white noise perturbation ξ_n to the dynamical system corresponds to random kicking the ball permitting thus to get out of the local minima. These kicks have to eventually vanish however since otherwise they could take the ball outside the basin of attraction of the global minimum and thus prevent the process from converging. This precise balance between the "downwards" driving term δ_n

and the random kicking term ϵ_n gives essentially the condition of convergence (see [30] for precise statements).

3.3.2. *Simulated annealing.* The stochastic minimisation algorithm known as *simulated annealing* proceeds [7, 41] by changing the *a priori* measure μ_0 in a way exponentially suppressing the sets of parameters leading to a large learning error [81]. This is achieved by introducing a positive parameter β, interpreted as the inverse temperature, and defining

$$\mu_{A,\beta}(d\theta) = \frac{\exp(-\beta \mathcal{H}_A(\theta))}{\mathcal{Z}_A(\beta)} \mu_0(d\theta),$$

where

$$\mathcal{Z}_A(\beta) = \int_\Theta \exp(-\beta \mathcal{H}_A(\theta)) \mu_0(d\theta).$$

Denote by $\Theta_{\min}^A = \{\theta \in \Theta : \mathcal{H}_A(\theta) = 0\} \subset \Theta$ the set composed from the network realisations minimising the total training error. When $\beta \to \infty$, it is intuitively clear that $\mu_{A,\beta}$ charges solely $\Theta_{\min}$. Of course, taking the limit $\beta \to \infty$ at this stage removes any computational advantage of the formalism since we recover the minimisation problem we had started with. Instead, it is much more efficient to use *simulating annealing algorithm* for attaining the minimum.

In mathematical terms, simulating annealing is the choice of a *cooling schedule* — *i.e.* a monotonically diverging sequence $(\beta_n)_{n \geq 1}$ of inverse temperatures $\beta_n \uparrow \infty$ — and the construction of an *irreducible inhomogeneous Markov chain* on Θ with transition probability kernel $p_{\beta_n}(\cdot, \cdot) : \Theta \times \mathcal{B}(\Theta) \to [0, 1]$, indexed by the sequence $(\beta_n)_{n \geq 1}$. The first requirement is that for fixed n, the measure μ_{A,β_n} must be an invariant measure for the Markov evolution *i.e.* it must be the left normalised eigenvector corresponding to the eigenvalue one for the Markov operator

$$\int_\Theta \mu_{A,\beta_n}(d\theta_1) p_{\beta_n}(\theta_1, \cdot) = \mu_{A,\beta_n}(\cdot).$$

This can be effectively implemented by standard algorithms like Metropolis, Kawasaki, *etc.* for any fixed β_n. The second requirement concerns the speed of cooling; provided that β_n diverges not very fast when $n \to \infty$, the inhomogeneous Markov chain, having transition kernel p_{β_n} at time n, converges to a measure

$\mu_{A,\infty}$ charging only the network realisations minimising the total training error, $\text{supp}\mu_{A,\infty} = \Theta^A_{\min}$. This result is precisely stated in the following

Theorem 3.8 (Simulated annealing). Let $(Y_n)_{n\geq 0}$ be an inhomogeneous irreducible Markov chain on Θ defined by the transition probability kernel p_{β_n},

$$\mathbb{P}(Y_{n+1} \in d\theta' | Y_n = \theta) = p_{\beta_n}(\theta, d\theta').$$

Then,

$$\lim_{n\to\infty} \mathbb{P}(Y_n \in \Theta^A_{\min}) = 1$$

if, and only if, the cooling schedule (β_n) verifies

$$\sum_{n=1}^{\infty} \exp(-\beta_n D) = \infty,$$

where D stands for the maximal depth of local minima of $\mathcal{H}_A$.

Proof. See [9]. ■

The previous theorem establishes the algorithmic implementability of such a minimisation scheme; moreover, it gives the optimal speed of the cooling schedules since the condition $\sum_{n=1}^{\infty} \exp(-\beta_n D) = \infty$ is satisfied provided $\lim \frac{\log n}{\beta_n} = c$ with $c \geq D$. In practice however such logarithmic cooling schedule proves very time consuming. To the extent of author's knowledge, *all* numerical results for large systems based on simulated annealing that are published in the literature are obtained with exponentially fast cooling schedules!

3.4. THERMODYNAMIC FORMALISM OF LEARNING

Beyond the practical algorithmic reasons, this (thermodynamic) formalism allows a thorough understanding of the learning procedure in terms of information theory.

To be more specific and to avoid unnecessary complications, assume that all the neurones are binary, all layers are finite, and, moreover, the parameter set is discrete [95, 101]. Now the finiteness of the sensor and motor layers together with the binary nature of the neurones imply that the set of all possible mappings

$\mathcal{M} = \{f : X_0 \to X_L\}$ is discrete and finite. The *a priori* measure μ_0 on Θ induces a measure on $\mathcal{M}$, denoted by the same symbol, by

$$\mu_0(f) = \int \mu_0(d\theta) \mathbb{1}_{\{F_\theta = f\}}, \quad \text{for} \quad f \in \mathcal{M}.$$

Given an arbitrary measure ν on Θ, define its *entropy* by

$$S(\nu) = -\sum_{f \in \mathcal{M}} \nu(f) \log \nu(f),$$

with the convention $0 \log 0 = 0$. It can be shown (exercise) that $0 \leq S(\nu) \leq \log \operatorname{card} \mathcal{M}$. As usual in information theory, the entropy of the measure ν can be interpreted [48] as the richness of ν or equivalently the *computational diversity* of the network architecture.

Similarly, given two arbitrary measures μ and ν with $\mu \ll \nu$ define the *relative entropy* by

$$S(\mu|\nu) = \sum_{f \in \mathcal{M}} \mu(f) \log \frac{\mu(f)}{\nu(f)}.$$

Denote finally by $\mathcal{G}_A(\beta)$ the Gibbs free energy, defined by

$$\mathcal{G}_A(\beta) = -\frac{1}{\beta} \log \mathcal{Z}_A(\beta).$$

A straightforward computation of the mean learning error leads to the formula

$$\frac{\partial(\beta \mathcal{G}_A(\beta))}{\partial \beta} = \mathbb{E}\mathcal{H}_A \equiv \int_\Theta \mathcal{H}_A(\theta) \mu_{A,\beta}(d\theta) = \sum_{f \in \mathcal{M}} \mu_{A,\beta}(f) \sum_{\alpha=1}^{A} d(y^\alpha, f(x^\alpha)),$$

identifying thus the average total training error with the internal energy of the corresponding thermodynamic system. Using the trivial observation that on the set $\{F_\theta = f\}$, the learning error reads $\mathcal{H}_A(\theta) = \sum_\alpha d(y^\alpha, f(x^\alpha))$, we obtain the standard relation of thermodynamics

$$\begin{aligned}
S_A &\equiv S(\mu_A|\mu_0) \\
&= \sum_{f\in\mathcal{M}} \mu_A(f) \log \frac{\mu_A(f)}{\mu_0(f)} \\
&= \sum_{f\in\mathcal{M}} \mu_A(f) \left[\log \int_\Theta \exp(-\beta\mathcal{H}_A(\theta))1\!\!1_{\{F_\theta=f\}}\mu_0(d\theta) - \log\mathcal{Z}_A - \log\mu_0(f)\right] \\
&= \sum_{f\in\mathcal{M}} \mu_A(f) \left[-\beta\sum_{\alpha=1}^{A} d(y^\alpha, f(x^\alpha)) + \log\mu_0(f) - \log\mu_0(f) - \log\mathcal{Z}_A\right] \\
&= -\beta\mathbb{E}\mathcal{H}_A + \beta\mathcal{G}_A \\
&= \beta(\mathcal{G}_A - \mathbb{E}\mathcal{H}_A).
\end{aligned}$$

Thermodynamically, this formula relates the entropy with the free and internal energies. From the informational point of view, the meaning of the last relation is also clear. First observe that $S_0 = 0$ and that $S_{A+1} \geq S_A$. Relative entropy being interpreted as information gain, this monotonicity implies that, starting from *tabula rasa*, learning new examples from the training set increases the information. Moreover, we compute

$$\frac{\partial S_A}{\partial \mathbb{E}H_A} = -\beta$$

meaning that information increases through a minimisation of the learning error.

The above arguments are strictly valid for finite networks and for training sets that remain small compared to the size of the net. It is intuitively clear that when the training set starts increasing without bounds some *saturation* must occur since, otherwise, an infinite quantity of information should be stored into a finite system! Several numerical studies have been done on this saturation phenomenon and heuristic computations based on replica trick or annealed approximation showed a clear transition from a memory regime to a saturation regime [54, 96]. However a rigorous mathematical treatment of this phenomenon is still an open problem.

4. Neural Network as Associative Memory

To fix ideas, we consider a McCulloch-Pitts network with N neurones at every layer and m particular fixed configurations of X_N, denoted $\xi^\mu = (\xi_1^\mu, \ldots, \xi_N^\mu)$ for

$\mu = 1, \ldots, m$, called *patterns*. We also use the symbol ξ_i to denote the family of variables $\xi_i = (\xi_1^\mu, \ldots, \xi_N^\mu)$ for $i = 1, \ldots, N$, so that upper indices number different patterns and lower indices number different sites.

4.1. CHOICE OF SYNAPTIC EFFICIENCES

We wish the net to memorise the patterns and to be able to recall them when *a clue* is presented to it. Now synaptic efficiencies J_{ij} must be *local*, *i.e.* they must depend only on ξ_i and ξ_j. This is a very important general principle; it excludes the possibility that the synapse connecting neurone i with neurone j needs some global information beyond that contained in ξ_i and ξ_j.

A reasonable and convenient additional assumption, but no so crucial as locality, is the *exchange-symmetry*, *i.e.* the constraint $J_{ij} = J_{ji}$. This allows actually to express the network evolution dynamics in terms of *Hamiltonian dynamics*. The most general form [43] of synaptic efficiencies satisfying locality and symmetry is in terms of a *symmetric synaptic kernel* $Q : \mathbb{R}^m \times \mathbb{R}^m \to \mathbb{R}$ by

$$J_{ij} = \frac{1}{N} Q(\xi_i; \xi_j), \quad \text{for} \quad i, j = 1, \ldots, N.$$

Several forms have been proposed for the synaptic kernel:

1. *Hebb's rule [42]:* the synaptic efficiencies are given by the formula

$$J_{ij} = \frac{1}{N} \sum_{\mu=1}^{m} \xi_i^\mu \xi_j^\mu = \frac{1}{N} \xi_i \cdot \xi_j,$$

   where $\cdot$ denotes the scalar product of $\mathbb{R}^m$.

2. *Clipped synapses:* we choose here

$$J_{ij} = \frac{1}{N} \phi(\xi_i \cdot \xi_j),$$

   where $\phi(x) = \text{sgn}(x)$.

3. *Truncated synapses:* the choice here is given by

$$J_{ij} = \frac{1}{N} \phi_a(\xi_i \cdot \xi_j),$$

where, for some positive constant $a \leq N$ the truncating function ϕ_a is defined by

$$\phi_a(x) = \begin{cases} a & \text{for } x > a \\ x & \text{for } |x| \leq a \\ -a & \text{for } x < -a. \end{cases}$$

4. *Inversion-symmetric synapses [43]:* for binary neurones, and for a given site i, the possible values of the vector ξ_i are the extremal points of the hypercube $[-1,1]^m$ of $\mathbb{R}^m$, namely the set $\{-1,1\}^m$. Denote by G_m the Abelian group generated by the m inversions

$$\begin{aligned} g_\mu \xi_i &= g_\mu(\xi_i^1, \ldots, \xi_i^{\mu-1}, \xi_i^\mu, \xi_i^{\mu+1}, \ldots, \xi_i^m) \\ &= (\xi_i^1, \ldots, \xi_i^{\mu-1}, -\xi_i^\mu, \xi_i^{\mu+1}, \ldots, \xi_i^m) \quad \text{for} \quad \mu = 1, \ldots, m. \end{aligned}$$

The group G_m contains 2^m elements and $g^2 = e$ for all $g \in G_m$. We say that a synaptic kernel is *totally inversion-symmetric* if for every $x, y \in \mathbb{R}^m$ and for every $g \in G_m$, it verifies

$$Q(gx, gy) = Q(x, y).$$

This symmetry provides a nice simplification since the 2^m characters of the group G_m are then eigenvectors of the kernel Q and can therefore serve as a basis for the spectral decomposition of Q from which interesting results can be easily obtained [43, 44, 29].

4.2. CHARACTERISATION OF PATTERNS

We are interested in obtaining *generic* results holding for a vast class of memorised patterns and not for some very specific ones in the limit when the network becomes very large: we don't expect the obtained results to hold for *all* patterns. We must therefore be able to characterise simply the set of patterns for which the net functions as an associative memory. This can be achieved by defining the patterns as random variables over some probability space $(\Omega, \mathcal{F}, \mathbb{P})$ and asking whether the results hold almost surely, or in the mean, or with large probability.

Several choices have been used to characterise the possible patterns: independent, correlated, or Gibbsian. In the sequel, we discuss briefly the various possibilities.

4.2.1. *Independent patterns.* The variables $(\xi_i^\mu)_{i=1,\dots,N}^{\mu=1,\dots,m}$ are independent, identically distributed random variables. For binary neurones we can use random variables with values in $\{-1,1\}$ but more general value-spaces do not alter the lines of reasoning. This choice has the advantage of being simple to realise and highly non trivial rigorous results can be obtained [13, 17, 28, 38, 49, 50, 61, 71, 83, 99, 103] ...

Epistemologically and philosophically it is however questionable whether this choice offers a good modelling of reality. On the other hand, several argue that if we are able to obtain results in that case, we can expect that the net will in fact perform better. Actually, the case of independent patterns is thought as the worst case; experience shows that it is much easier for the human brain to recall a well structured information than a collection of unrelated random facts. This anthropic way of thinking is however unreliable since in the absence of any closed theory of neural computing, it is not granted that neural nets perform in the same way as human brain: this is the fact we wish to establish. Epistemologically and logically it is not acceptable to include such an hypothesis as postulate.

Nevertheless, it is worth exploring the case of independent random variables because we can hope obtaining a mathematically closed theory within a reasonably remote future that could serve as a first approximation of a more general and realistic theory.

4.2.2. *Spatially or semantically correlated patterns.* At the neurophysiological level, many examples of *spatial correlation* are known in the context of visual information processing [27]. *Semantic correlations*, on the other hand, occur when patterns are splitted into categories and subcategories for the purpose of classification and biological evidence of this phenomenon is given in [68]. Therefore, an intensive effort to model such phenomena was made [40, 69, 100].

For our purposes, instead of considering patterns that are independent and identically distributed symmetric random variables with variance one, we consider families of random variables $(\xi_i^\mu)_{i=1,\dots,N}^{\mu=1,\dots,m}$ over an underlying probability space $(\Omega, \mathcal{F}, \mathbb{P})$ with a more complicated dependence structure. We always consider variables with $\mathbb{E}\xi_i^\mu = 0$ and $\mathbb{E}[(\xi_i^\mu)^2] = 1$. In the case of spatially correlated patterns, we impose

$$\mathbb{E}(\xi_i^\mu \xi_j^\nu) = \delta^{\mu\nu} C_{ij},$$

where $(C_{ij})_{ij}$ is a $N \times N$ non-trivial matrix of spatial correlations. For semantic correlations, we impose

$$\mathbb{E}(\xi_i^\mu \xi_j^\nu) = \delta_{ij} \widetilde{C}^{\mu\nu},$$

where $(\widetilde{C}^{\mu\nu})^{\mu\nu}$ is a $m \times m$ non-trivial matrix of semantic correlations.

Several interesting, but not rigorous, results concerning the storage capacity of such models are obtained in [69, 100] where, by use of replica trick, it is shown that the critical capacity for storing correlated patterns exceeds the capacity of storing independent ones. This is a very appealing characteristic of the replica trick calculations and tends to moderate the somehow severe epistemological criticism made in the previous subsesction. It should be interesting to be able to obtain rigorous results for such correlated patterns.

4.2.3. *Gibbsian patterns.* The next step is to introduce mixed semantic and spatial correlations. To keep the models tractable, some particular form of correlation must be chosen. In that direction, Schlüter and Wagner [92] introduced Gibbsian structure for the patterns. More specifically, the vertex set indexing the configurations (hence the patterns) is now a finite subset, Λ_n, of the D-dimensional lattice $\mathbb{Z}^D$, *i.e.* we have a family of random variables $(\xi_i^\mu)_{i\in\Lambda_n}^{\mu=1,\dots,m}$, over an underlying probability space $(\Omega, \mathcal{F}, \mathbb{P})$, taking values in $\{-1, 1\}$, with $\Lambda_n = [-n, n]^D \cap \mathbb{Z}^D$. We denote by κ the product measure

$$\kappa = (\delta_{-1} + \delta_1)^{\mathbb{Z}^D}$$

on $\{-1, 1\}^{\mathbb{Z}^D}$ and impose a Gibbsian distribution to the random variables ξ given by

$$g_{n,m,\lambda}(\xi \in d\eta) = \frac{\exp\left(\frac{\lambda}{\sqrt{m}} \sum_{\mu=1}^{m} \sum_{\langle i,j\rangle} \eta_i^\mu \eta_j^\mu\right)}{z_m(\lambda)} \prod_{i\in\Lambda_n} \prod_{\mu=1}^{m} \kappa(d\eta_i^\mu),$$

and

$$z_m(\lambda) = \int_{\{-1,1\}^{m|\Lambda_n|}} \exp\left(\frac{\lambda}{\sqrt{m}} \sum_{\mu=1}^{m} \sum_{\langle i,j \rangle} \eta_i^\mu \eta_j^\mu\right) \prod_{i \in \Lambda_n} \prod_{\mu=1}^{m} \kappa(d\eta_i^\mu),$$

where $\langle i,j \rangle$ means that the sum extends over pairs i,j of sites of Λ_n that are nearest neighbours.

In a very interesting pre-print, Nishimori, Whyte, and Sherrington [73] study the phase diagram of a short-range finite-dimensional network having the short-range 'Hebbian' form synaptic efficiencies, namely

$$J_{ij} = \begin{cases} \frac{1}{N} \sum_{\mu=1}^{m} \xi_i^\mu \xi_j^\mu & \text{if } |i-j| = 1 \\ 0 & \text{otherwise,} \end{cases}$$

in which the memorised patterns are distributed according to the Gibbs measure $g_{n,m,\lambda}$.

4.3. SEQUENTIAL HAMILTONIAN DYNAMICS FOR THE PASTUR-FIGOTIN-HOPFIELD NETWORK

Among the neural networks, a special *rôle* is played by the Pastur-Figotin (improperly called Hopfield) network [75, 76, 46]. Consider a McCulloch and Pitts network with N binary neurones at every layer and consider m particular fixed configurations of X_N, denoted $\xi^\mu = (\xi_1^\mu, \ldots, \xi_N^\mu)$ for $\mu = 1, \ldots, m$. These configurations are called *patterns*. Fix the synaptic efficiencies by the Hebb's rule [42]

$$J_{ij} = \frac{1}{N} \sum_{\mu=1}^{m} \xi_i^\mu \xi_j^\mu$$

and let all the threshold w_i vanish. The evolution of the network proceeds as usual, through[5]

$$x_i(t+1) = \operatorname{sgn}\left(\frac{1}{N} \sum_{\mu=1}^{m} \sum_{1 \le i \neq j \le N} \xi_i^\mu \xi_j^\mu x_j(t)\right).$$

[5] Notice that the factor $\frac{1}{N}$ in this formula is completely irrelevant as far as evolution is concerned. It is there merely for later thermodynamical considerations.

For reasons that will become clear in the next section, we also introduce the so called *Hopfield Hamiltonian*

$$H_N(x) = \frac{2}{N}\sum_{\mu=1}^{m}\sum_{1\le i<j\le N} \xi_i^\mu \xi_j^\mu x_i x_j.$$

It is immediate to verify that if $m = 1$, the pattern $\xi = \xi^1 = (\xi_1^1, \cdots, \xi_N^1)$ is a fixed point of the network dynamics (check!). What is more important however is the functioning of the network as an associative memory; consider in fact the simplest case $m = 1$, and an initial configuration

$$x_i(0) = \begin{cases} \xi_i & \text{for } i \in A \\ -\xi_i & \text{for } i \notin A, \end{cases}$$

where $A \subseteq \{1, \ldots, N\}$ with card $A < \frac{N}{2}$.

It is immediate to show that $x_i(1) = \xi_i$, $\forall i$, recovering thus the memorised pattern ξ, by presenting to the network a clue, even substantially differing from the memorised pattern. The situation is of course more complicated in the case of $m > 1$ but proceeds essentially in the same lines. Neural networks being usually very large, important effort is made in the understanding the limit $N\to\infty$ and $m\to\infty$ simultaneously. It turns out that when the number of patterns grows too rapidly with respect to the size of the net, a memory saturation phenomenon appears and the network is unable to recall memorised patterns any more.

The most important results, summarised in the sequel, are obtained for a family (ξ_i^μ) of independent Bernoulli variables.

4.3.1. *Asymptotic stability and attraction.* We consider a Hopfield network whose size and number of memorised patterns tend eventually to infinity. We have the following

Theorem 4.1. Let $m = \frac{N}{\gamma \log N}$. Then

1. if $\gamma > 6$, then, for $N\to\infty$, the m original patterns are almost surely stable, *i.e.*

$$\mathbb{P}[\liminf_N (\cap_{\nu=1}^m \{T\xi^\nu = \xi^\nu\})] = 1$$

2. if $\gamma > 4$, then

$$\mathbb{P}[\cap_{\nu=1}^{m}\{T\xi^{\nu} = \xi^{\nu}\}] = 1 - R_N,$$

with $\lim_{N\to\infty} R_N = 0$.

Proof. Fix some pattern ν. Updating the i-th component, involves the estimation of

$$\operatorname{sgn}\left(\sum_{j\neq i}\xi_i^{\nu}(\xi_j^{\nu})^2 + \sum_{\mu\neq\nu}\xi_i^{\mu}\sum_{j\neq i}\xi_j^{\mu}\xi_j^{\nu}\right) = \xi_i^{\nu}\operatorname{sgn}\left(N - 1 + \xi_i^{\nu}\sum_{\mu\neq\nu}\xi_i^{\mu}\sum_{j\neq i}\xi_j^{\mu}\xi_j^{\nu}\right).$$

Therefore the pattern ξ^{ν} is stable over the site i on the event $A^c(N, m, i, \nu)$, where

$$A(N, m, i, \nu) = \left\{N - 1 + \xi_i^{\nu}\sum_{\mu\neq\nu}\xi_i^{\mu}\sum_{j\neq i}\xi_j^{\mu}\xi_j^{\nu} \leq 0\right\}.$$

Random variables ξ_i^{μ} being bounded, they posses exponential moments; hence the result is obtained by use of Markov inequality. Notice that

$$\begin{aligned}\mathbb{P}[\cup_{\nu=1}^{m}\cup_{i=1}^{N} A(N, m, i, \nu)] &\leq \sum_{\nu=1}^{m}\sum_{i=1}\mathbb{P}[A(N, m, i, \nu)] \\ &\leq \sum_{\nu=1}^{m}\sum_{i=1}^{N}\inf_{t>0}\exp[-t(N-1)]\mathbb{E}\exp\left(-t\xi_i^{\nu}\sum_{\mu\neq\nu}\xi_i^{\mu}\sum_{j\neq i}\xi_j^{\mu}\xi_j^{\nu}\right).\end{aligned}$$

Computing of the expectation is performed inductively by first computing conditional expectation with respect to σ-fields generated from adequate subsets of random variables ξ. Introduce in fact the following σ-fields:

$$\mathcal{F}_j^{\mu} = \sigma\left\{\left(\xi_k^{\lambda}\right)_{k=1,\cdots,N}^{\lambda=1,\cdots,m;\lambda\neq\mu}\right\}$$

and

$$\mathcal{F}_j^{\mu} = \sigma\left\{\left(\xi_k^{\lambda}\right)_{k=1,\cdots,N}^{\lambda=1,\cdots,m;\lambda\neq\mu};(\xi_k^{\mu})_{k=1,\cdots,N;k\neq j}\right\}$$

We get

$$
\begin{aligned}
\mathbb{E}\exp\left(-t\xi_i^\nu \sum_{\mu\neq\nu}\xi_i^\mu \sum_{j\neq i}\xi_j^\mu\xi_j^\nu\right) &= \mathbb{E}\left(\mathbb{E}(\prod_{\mu\neq\nu}\exp(-t\xi_i^\nu\xi_i^\mu\sum_{j\neq i}\xi_j^\mu\xi_j^\nu)|\mathcal{F}^\nu)\right)\\
&= \mathbb{E}\left(\prod_{\mu\neq\nu}\mathbb{E}(\exp(-t\xi_i^\nu\xi_i^\mu\sum_{j\neq i}\xi_j^\mu\xi_j^\nu)|\mathcal{F}^\nu)\right)\\
&= \mathbb{E}\left(\prod_{\mu\neq\nu}\mathbb{E}\left(\mathbb{E}(\prod_{j\neq i}\exp(-t\xi_i^\nu\xi_i^\mu\ \xi_j^\mu\xi_j^\nu)|\mathcal{F}_i^\nu)|\mathcal{F}^\nu\right)\right)\\
&= \mathbb{E}\left(\prod_{\mu\neq\nu}\mathbb{E}\left(\prod_{j\neq i}\mathbb{E}(\exp(-t\xi_i^\nu\xi_i^\mu\ \xi_j^\mu\xi_j^\nu)|\mathcal{F}_i^\nu)|\mathcal{F}^\nu\right)\right)\\
&= (\cosh t)^{(N-1)(m-1)}.
\end{aligned}
$$

Now $\cosh t \leq \exp(t^2/2)$; hence

$$
\begin{aligned}
\mathbb{P}\left[\cup_{\nu=1}^m \cup_{i=1}^N A(N,m,i,\nu)\right] &\leq \sum_{\nu=1}^m\sum_{i=1}^N \mathbb{P}[A(N,m,i,\nu)]\\
&\leq \sum_{\nu=1}^m\sum_{i=1}^N \inf_{t>0}\exp[-t(N-1)]\exp\left[\frac{t^2}{2}(N-1)(m-1)\right]\\
&= mN\inf_{t>0}\exp\left[-t(N-1)+\frac{t^2}{2}(N-1)(m-1)\right]\\
&= mN\exp\left(-\frac{N-1}{2(m-1)}\right).
\end{aligned}
$$

The first part of the proof is completed by use of Borel-Cantelli lemma since the choice of $\gamma > 6$ leads to a convergent series $\sum_{N\geq 2}\frac{1}{N(\log N)^2}$. For the weak part of this theorem to be true, it is enough that $\lim_{N\to\infty} mN\exp(-\frac{N-1}{2(m-1)}) = 0$; this happens in fact for $\gamma > 4$. ■

The weak part of the previous theorem was formulated in [62] and proved in [64] and together with the almost sure version, in [103]; it guarantees that the original patterns are stable under sequential evolution, provided that their number m grows slightly sub linearly in N. This stability is of course a necessary condition for the network to behave as an associative memory. However, attraction of configurations towards stable patterns is also needed. This is provided by the following

Theorem 4.2. Let $\rho \in [0, 1/2[$ and $m = (1-2\rho)^2 \frac{N}{\gamma \log N}$. For $\nu = 1, \ldots, m$, let $x^\nu(0)$ be a configuration on the Hamming sphere centred at ξ^ν and of radius ρN. Then,

1. if $\gamma > 6$,
$$\mathbb{P}[\liminf_N (\cap_{\nu=1}^m \{Tx^\nu(0) = \xi^\nu\})] = 1$$

2. if $\gamma > 4$,
$$\mathbb{P}[\cap_{\nu=1}^m \{Tx^\nu(0) = \xi^\nu\}] = 1 - R_N,$$
with $\lim_{N\to\infty} R_N = 0$.

Proof. The proof is developed along the same lines as the proof of the previous theorem by remarking that

$$\mathbb{P}[\{(Tx^\nu(0))_i = -\xi_i^\nu\}] \le \exp\left[-\frac{(N(1-2\rho)-1)^2}{(m-1)(N-1)}\right] \equiv K_{N,m}(\rho)$$

where $K_{N,m}(\rho)$ is an increasing function of ρ. Observe also that

$$\mathbb{P}[\cup_{\nu=1}^m \{Tx^\nu(0) = \xi^\nu\}] \le mNK_{N,m}(\rho)$$

and conclude by optimising on the parameters m for the Borel-Cantelli lemma to apply. ■

Remark. Results of the previous kind were first established in [52] and a complete proof was given in [103]. The natural question one can ask is whether *all* the vectors of the Hamming spheres $\cup_{\nu=1}^m S(\xi^\nu, \rho N)$ *do* converge directly to the original patterns. It is possible to choose some particular vectors, containing only $\mathcal{O}(\sqrt{N})$ errors that do not converge to the original pattern, providing thus a negative answer to the previous question. Notice however that these vectors are explicitly constructed not to converge and hence they are not in contradiction with the almost sure result stated in the theorem.

It is quite remarkable that these results can be extended to other initial configurations, namely random mosaics. These are configurations coinciding with a given original pattern on some sites, with a different original pattern on some other sites and so on. It is even allowed to have an independent random noise on some sites. A theorem of attraction similar to the previous one can be shown for such random mosaics [104]. Namely, it is shown the

Theorem 4.3. Let $(\xi_i^{m+1})_{i=1,\cdots,N}$ be a collection of N random variables, mutually independent and independent of the original variables $(\xi_i^\mu)_{i=1,\cdots,N}^{\mu=1,\cdots,m}$ having the same distribution with the original variables. Let $n_0, n_1, \cdots, n_{m+1}$ be an increasing family of positive integers with $0 = n_0 \le n_1 \le \cdots \le n_{m+1} = N$ and $(I_\lambda)_{\lambda=1,\cdots,m+1}$ be a partition of the set of indices $\{1, \cdots, N\}$ with $\mathrm{card}(I_\lambda) = n_\lambda - n_{\lambda-1}$. Let

$$x^\nu(0) = \begin{cases} \xi_i^\nu & \text{for } i \in I_1 \\ \xi_i^{a^\nu(\mu)} & \text{for } i \in I_\mu, \quad \mu = 2, \cdots, m \\ \xi_i^{m+1} & \text{for } i \in I_{m+1}, \end{cases}$$

where a^ν is a transposition of the index set $\{1, \cdots, N\}$ with $a^\nu(1) = \nu$ and $a^\nu(\nu) = 1$. Let $m = \frac{N}{\gamma \log N}$, $n_1 = \gamma_1 N$, and $n_m = \gamma_m N$, with $\gamma > 0$ and $0 < \gamma_1 \le \gamma_m \le 1$ and let finally ϵ be arbitrarily small. Then

1. If $\gamma_1 > \frac{1}{2}(\gamma_m + \epsilon)$ and $\gamma > \frac{6}{(2\gamma_1 - \gamma_m - \epsilon)^2}$ then

$$\mathbb{P}\left(\liminf_N [\cap_{\nu=1}^m \{Tx^\nu(0) = \xi^\nu\}]\right) = 1.$$

2. If $\gamma_1 > \frac{1}{2}(\gamma_m + \epsilon)$ and $\gamma > \frac{4}{(2\gamma_1 - \gamma_m - \epsilon)^2}$ then

$$\mathbb{P}\left(\cap_{\nu=1}^m \{Tx^\nu(0) = \xi^\nu\}\right) = 1 - R_N,$$

with $\lim_{N\to\infty} R_N = 0$.

However, all these are only partial results. It should be interesting to be able to prove attraction after a long number of time steps, or even asymptotically in time, but this remains still a challenging open problem.

4.3.2. *Fluctuations of the Hamiltonian and existence of local minima.* It can easily be shown that the Hopfield Hamiltonian decreases in time under the network evolution (exercise !). Moreover, the stable configurations are local minima of the Hamiltonian. It is therefore interesting to study the structure of local minima. A first result in this direction has been proven by Newman in 1988 by use of large deviation techniques [71]. Namely, he showed the following

Theorem 4.4. There exists a number $\alpha_c > 0$ such that for all $\alpha \leq \alpha_c$ and $m = \alpha N$, there exist $\delta \in]0, 1/2[$ and $\epsilon > 0$ such that

$$\mathbb{P}[\liminf_N (\cap_{\nu=1}^m \cap_{y \in \mathcal{S}(\xi^\mu, \delta N)} \{H_{N,m}(y) > H_{N,m}(\xi^\mu) + \epsilon N\})] = 1$$

where $\mathcal{S}(\xi^\mu, \delta N)$ is the Hamming sphere centred at ξ^μ and of radius δN.

Moreover, Newman obtained the numerical value $\alpha_c > 0.056$. On the other hand, Amit and his collaborators [5, 6], based on numerical simulations and non rigorous computations, predict $\alpha_c = 0.14$. In a more recent work [60, 61], Loukianova obtained $\alpha_c > 0.071$; she used large deviation techniques similar to that of Newman, but a finer decomposition of the space and non-uniform estimates. We give here this more recent proof.

Proof. Denote by $\mathcal{S}(x, r)$ the sphere centred at the configuration x with (Hamming) radius r and define $H_N(x, r) = \min_{x' \in \mathcal{S}(x,r)} H_N(x')$.

For $\delta \in]0, 1/2[$ and $\epsilon > 0$, denote by $A(N, m, \delta, \epsilon)$ the event that each pattern is surrounded by an energy barrier of height ϵN on the sphere of radius $[\delta N]$:

$$A(N, m, \delta, \epsilon) = \cap_{\mu=1}^m \{h_N(\xi^\mu, [\delta N]) > H_N(\xi^\mu) + \epsilon N\}.$$

Put $M = m - 1$ and define the random variables

$$W_N^\mu = \frac{1}{\sqrt{N}} (\xi^\mu, \xi^m).$$

These random variables are independent and identically distributed; their law weakly converges, when $N \to \infty$, towards a standard normal law. Define, for $\eta > 0$,

$$\Omega_\eta = \left\{ \frac{1}{M} \sum_{\mu=1}^M (W_N^\mu)^2 \in]1 - \eta, 1 + \eta[\right\}.$$

Obviously,

$$\mathbb{P}[A^c(N,m,\delta,\epsilon)] \leq \mathbb{P}(\Omega_\eta^c) + \mathbb{P}[A^c(N,m,\delta,\epsilon)\cap\Omega_\eta].$$

For $J \subseteq \{1,\ldots,N\}$ and $x \in X_N$, denote by x_J the configuration that differs from x exactly on the co-ordinates J. It is immediate then to see that $x_J \in \mathcal{S}(x,|J|)$. Thus

$$\begin{aligned}\mathbb{P}[A^c(N,m,\delta,\epsilon)\cap\Omega_\eta] &= \mathbb{P}[\cup_\mu \cup_{J:|J|=[\delta N]} \{H(\xi_J^\mu) - H(\xi^\mu) \leq \epsilon N\} \cap \Omega_\eta] \\ &\leq \sum_\mu \sum_{J:|J|=[\delta N]} \mathbb{P}[\{H(\xi_J^\mu) - H(\xi^\mu) \leq \epsilon N\}\cap\Omega_\eta] \\ &\leq mC_N^{[\delta N]}\mathbb{P}[\{H(\xi^m_{\{1,\ldots,[\delta N]\}}) - H(\xi^m) \leq \epsilon N\}\cap\Omega_\eta].\end{aligned}$$

Now remark that the variables appearing in the last event can be expressed in terms of the variables W^μ by

$$N\Big(H(\xi^m_{\{1,\ldots,[\delta N]\}}) - H(\xi^m)\Big) = 4[\delta N]\,(N-[\delta N]) + 4\sqrt{[\delta N](N-[\delta N])}\sum_{\mu=1}^{M}\overline{W}_N^\mu\widetilde{W}_N^\mu,$$

where

$$\overline{W}_N^\mu = \frac{1}{\sqrt{[\delta N]}}\sum_{i=1}^{[\delta N]}\xi_i^\mu\xi_i^m$$

and

$$\widetilde{W}_N^\mu = \frac{1}{\sqrt{N-[\delta N]}}\sum_{i=1+[\delta N]}^{N}\xi_i^\mu\xi_i^m.$$

Introduce the quantities

$$q_N(M,\eta,C) = \mathbb{P}\left[\left\{\frac{1}{M}\sum_{\mu=1}^{M}\overline{W}_N^\mu\widetilde{W}_N^\mu < C\right\}\cap\Omega_\eta\right]$$

and

$$p_N(M,\eta) = \mathbb{P}(\Omega_\eta^c)$$

to obtain

$$\mathbb{P}[A^c(N,m,\delta,\epsilon)] \leq p_N(M,\eta) + C_N^{[\delta N]}q_N\left(M,\eta,\frac{-\sqrt{\delta(1-\delta)}}{\alpha}+\epsilon'\right),$$

for all $\epsilon' > \frac{\epsilon}{4\alpha\sqrt{\delta(1-\delta)}}$ and N sufficiently large.

Let

$$\Lambda(t,r) = \lim_{M\to\infty} \frac{1}{M} \log \mathbb{E} \exp \left\{ t \sum_{\mu=1}^{M} (W_N^\mu)^2 + r \sum_{\mu=1}^{M} \overline{W}_N^\mu \widetilde{W}_N^\mu \right\}$$

and

$$\Lambda^*(x,y) = \sup_{t,r} (xt + yr - \Lambda(t,r))$$

its Legendre dual. We have then the standard large deviation estimate

$$\limsup_N \frac{1}{N} \log q_N(M, \eta, \frac{-\sqrt{\delta(1-\delta)}}{\alpha} + \epsilon') \leq - \inf_{1-\eta\leq x\leq 1+\eta} \alpha\Lambda^*(x, \frac{-\sqrt{\delta(1-\delta)}}{\alpha} + \epsilon')$$

and

$$\limsup_N \frac{1}{N} \log p_N(M, \eta) \leq -\frac{\alpha}{2}(\eta - \log(1+\eta)).$$

Denoting $\mathcal{I}(\delta) = -\delta \log \delta - (1-\delta)\log(1-\delta)$, Stirling formula guarantees that $\lim \frac{1}{N} \log C_N^{[\delta N]} = \mathcal{I}(\delta)$.

Now $\limsup_N \frac{1}{N} \log \mathbb{P}[A^c(N, m, \delta, \epsilon)] < -K(\alpha, \delta, \epsilon)$, with $K(\alpha, \delta, \epsilon) > 0$ provided that we can choose $\alpha \in]0, \infty[$ and $\delta \in]0, 1/2[$ such that, for $\eta > 0$ and $\epsilon > 0$, we have simultaneously

$$-\mathcal{I}(\delta) + \inf_{1-\eta\leq x\leq 1+\eta} \alpha\Lambda^* \left(x, \frac{-\sqrt{\delta(1-\delta)}}{\alpha} + \epsilon' \right) > 0$$

and

$$\eta - \log(1+\eta) > 0.$$

It is shown in [61] that this is possible; numerical estimation gives then $\alpha^* \simeq 0.0712$ and for the corresponding $\delta \simeq 0.009$. ■

Remark. The estimate for α^* has been further improved in [99] to the value $\alpha^* \simeq 0.085$. Obtaining rigorously a finite upper bound of α_c still remains an open problem.

Another result along the same lines was also proved in [61]

Theorem 4.5. For given integers N and m, and $\delta > 0$, let

$$C(N, m\delta) = \{\exists \mu \in \{1, \cdots, m\} : \exists x \in \mathcal{B}(\xi^\mu, [\delta N]) \text{ such that } x \text{ is a local minimum of } H\},$$

where $\mathcal{B}(\xi^\mu, [\delta N])$ is the Hamming ball around the configuration ξ^μ of radius $[\delta N]$, the symbol $[\cdot]$ denoting the integer part. Suppose that $\lim_{N\to\infty} \frac{m}{N} = \alpha > 0$. Then there exists a positive constant $\delta(\alpha)$, depending on α, such that

$$\liminf_{N\to\infty} \frac{1}{N} \log \mathbb{P}[C(N, m, \delta)] \leq -K(\alpha, \delta)]$$

with $K(\alpha, \delta) > 0$ and $\liminf_{\alpha\to\infty} \delta(\alpha) \geq \delta_* \simeq 0.05$.

Remark. Notice that the set $C(N, m\delta)$ can be rewritten as

$$C(N, m\delta) = \cup_{\mu=1}^{m} \cup_{k=1}^{[\delta N]} \cup_{\substack{J \subseteq \{1, \cdots, N\} \\ |J|=k}} \{\xi_J^\mu \text{ is a local minimum of } H\},$$

where

$$\{\xi_J^\mu \text{ is a local minimum of } H\} = \{\forall i \in J : H(\xi_{J\setminus\{i\}}^\mu) > H(\xi_J^\mu)\} \cap \{\forall i \notin J : H(\xi_{J\cup\{i\}}^\mu) > H(\xi_J^\mu)\}.$$

Thus the previous theorem establishes that there are no local minima in small balls around each pattern.

With a view to numerical simulations, it is very interesting to know whether near a given configuration $x \in X_N$ there are many Hamming neighbours having lower energy than x. This question can be formalised more precisely in terms of the notion of escape direction. For a given configuration x and a given site i, denote as usual by $x_{\{i\}}$ the configuration that differs from x only at site i. Thus $x_{\{i\}}$ is a nearest Hamming neighbour of x.

Definition 4.6. A site $i \in \{1, \ldots, N\}$ is an *escape direction* for the configuration x if

$$H_{N,m}(x) > H_{N,m}(x_{\{i\}}).$$

We denote by

$$\mathcal{N}_N(x) = \text{card}\{i \in \{1,\ldots,N\} : i \text{ is an escape direction of } x\}.$$

It can then be proved [61] the following

Theorem 4.7. Let $\lim \frac{m}{N} = \alpha > 0$. There exist two strictly positive parameters $\gamma_*(\alpha)$ and $\gamma^*(\alpha)$ such that, for every $\mu = 1,\ldots,m$,

$$\gamma_*(\alpha) \leq \liminf_N \frac{\mathcal{N}_N(\xi^\mu)}{N} \leq \limsup_N \frac{\mathcal{N}_N(\xi^\mu)}{N} \leq \gamma^*(\alpha).$$

Numerical values for the parameters $\gamma_*(\alpha)$ and $\gamma^*(\alpha)$ can be explicitly obtained as functions of α. It is argued that the critical storage capacity should have an interpretation also in terms of the asymptotic number of escape directions.

5. Thermodynamics of the Restitution Process

5.1. MAIN DEFINITIONS

We again limit ourselves to the Hopfield model of binary neurones, sequentially updated. Now however, we allow for a stochastic evolution, *i.e.* instead of rigidly imposing the evolution

$$x_i(t+1) = \text{sgn}\left(\frac{1}{N}\sum_{\mu=1}^{m}\sum_{1\leq i<j\leq N} \xi_i^\mu \xi_j^\mu x_j(t)\right),$$

we introduce the $2^N \times 2^N$ transition matrices

$$p_i^{(\beta)}(x,y) = \begin{cases} \frac{\exp(\beta h_i(x) y_i)}{\exp(\beta h_i(x)) + \exp(-\beta h_i(x))} & \text{if } y_j = x_j \text{ for } j \neq i \\ 0 & \text{otherwise,} \end{cases}$$

for $i = 1,\ldots,N$, and $x, y \in \{-1,1\}^N$ where

$$h_i(x) = \frac{1}{N}\sum_{\mu=1}^{m}\sum_{j:j\neq i} \xi_i^\mu \xi_j^\mu x_j.$$

These transition matrices allow the definition of a Markov chain $(Y_n)_{n\geq 0}$ on $\{-1,1\}^N$ by

$$\mathbb{P}(Y_{n+1}=y|Y_n=x)=p^{(\beta)}(x,y),$$

where

$$p^{(\beta)}(x,y)=\sum_{z_1,\dots,z_N} p_1^{(\beta)}(x,z_1)\cdots p_N^{(\beta)}(z_N,y).$$

As it was the case for the stochastic learning procedure, the stochastic evolution tends to the deterministic one if one let $\beta\to\infty$. Gibbs formalism arises now naturally. Define the finite volume Gibbs measure as a probability on the configuration space given by

$$\gamma_{N,m,\beta}(\cdot)=\frac{1}{Z_{N,m,\beta}}\sum_x \exp(-\beta H_{N,m}(x))\mathbb{1}_{(\cdot)}(x),$$

where $Z_{N,m,\beta}$ is the partition function.

It is then easily shown the following

Proposition 5.1. The finite volume Gibbs measure $\gamma_{N,m,\beta}$ is the unique invariant measure associated with the Markov chain of transition probability p.

Proof. Exercise! ■

The specific free energy is defined, as usual, by

$$f_{N,m,\beta}=-\frac{1}{\beta N}\log Z_{N,m,\beta}$$

and since the Hamiltonian depends on the patterns ξ, so does the partition function and hence the free energy.

To obtain generic results, we can again choose the variables ξ to be Bernoulli independent random variables. Therefore the thermodynamic functions in finite volume are random variables as it is the case in spin glass systems [82]. The same questions can be asked, namely whether the thermodynamic variables tend to a non random limit when $N\to\infty$, or whether quenched, $f_{N,m,\beta}$, and annealed,

$$\overline{f}_{N,m,\beta}=-\frac{1}{\beta N}\log \mathbb{E}Z_{N,m,\beta},$$

free energies coincide in the infinite volume limit. These questions are partially answered in the following sections.

5.2. BEHAVIOUR OF THE ANNEALED FREE ENERGY

The asymptotic behaviour of the annealed free energy is established in the high and low temperature regimes [103].

Theorem 5.2. Let $\alpha = \lim \frac{m}{N}$. Then

1. if $\beta < 1$,

$$\lim_{\substack{N\to\infty \\ \alpha\geq 0}} \bar{f}_{N,m,\beta} = -\frac{\log 2}{\beta} + \frac{\alpha}{2\beta}\log(1-\beta) + \frac{\alpha}{2}$$

2. if $\beta > 1$,

$$\lim_{m,N\to\infty} \bar{f}_{N,m,\beta} = -\infty$$

and

$$\lim_{\substack{N\to\infty \\ m \text{ finite}}} \bar{f}_{N,m,\beta} = -\frac{\log 2}{\beta} + \frac{m}{\beta}\inf_{x\in\mathbb{R}}\left\{\frac{x^2}{2} - \log\cosh(\sqrt{\beta}x)\right\}.$$

Proof. The main idea of the proof is to introduce auxiliary Gaussian variables to linearise the quadratic form in the exponent of the partition function

$$Z_{N,m,\beta} = \exp\left(-\frac{\beta m}{2}\right)\sum_{x\in X_N}\prod_{\mu=1}^{m}\exp\left(\frac{\beta}{2N}(\xi^\mu, x)_N^2\right)$$

where $(\cdot,\cdot)_k$ denotes the scalar product of $\mathbb{R}^k$. Using the identity

$$\exp\left(\frac{x^2}{2a}\right) = \sqrt{\frac{a}{2\pi}}\int_{\mathbb{R}}\exp\left(-\frac{a}{2}t^2 + tx\right)dt,$$

we write

$$Z_{N,m,\beta} =$$

$$\exp(-\frac{\beta m}{2})(\frac{N}{2\pi})^{m/2}\int_{\mathbb{R}^m} dr_1\cdots dr_m \sum_{x\in X_N}\exp(-\frac{N}{2}\sum_{\mu=1}^{m} r_\mu^2 + \sqrt{\beta}\sum_{\mu=1}^{m} r_\mu\sum_{i=1}^{N} x_i\xi_i^\mu).$$

But now the integration over x is trivially performed and by Fubini's theorem we obtain

$$\mathbb{E}Z_{N,m,\beta} = 2^N \exp(-\frac{\beta m}{2}) \left[(\frac{N}{2\pi})^{1/2} \int_{\mathbb{R}} dr \exp(-\frac{Nr^2}{2} + N \log\cosh(\sqrt{\beta} r)) \right]^m .$$

Using standard Laplace's method, we obtain, asymptotically in N,

$$\overline{f}_{N,m,\beta} = -\frac{\log 2}{\beta} + \frac{m}{2N} - \frac{m}{2\beta N} - \frac{m}{2\beta N} \log\frac{N}{2\pi} - \frac{m}{\beta} \inf_{x\in\mathbb{R}} \{\frac{x^2}{2} - \log\cosh(\sqrt{\beta} x)\} + R_N,$$

with $\lim_N R_N = 0$.

Now, for $\beta > 1$

$$\inf_{x\in\mathbb{R}} \left\{ \frac{x^2}{2} - \log\cosh(\sqrt{\beta} x) \right\} = c_\beta < 0$$

and when $\lim_N m = \infty$ the annealed free energy diverges. On the contrary, for $\beta < 1$,

$$\inf_{x\in\mathbb{R}} \left\{ \frac{x^2}{2} - \log\cosh(\sqrt{\beta} x) \right\} = 0.$$

We split then the term $-\frac{Nr^2}{2}$ appearing in the exponential into $-\frac{Nr^2(1-\beta)}{2} - \frac{Nr^2\beta}{2}$, absorb $(1-\beta)$ in the integration variable, carry out the integration, and conclude. ■

5.3. BEHAVIOUR OF THE QUENCHED FREE ENERGY

The next point is to establish the existence of the limit of the quenched free energy. This can be proved only for a region of the parameters.

Theeorem 5.3. Let $\alpha \in [0,1[$ and $\delta \in]0,1[$ such that $\delta - 4\sqrt{\alpha}(1-\delta) > 0$. Let

$$\Phi_\delta :]0,\infty[\times \mathbb{R} \to \mathbb{R}$$
$$(\beta, h) \mapsto -\frac{1}{\beta} \log\cosh(\beta h) + \frac{1-\delta}{2} h^2,$$

and define two functions

$$\Psi^1(\beta,\alpha,\delta) = \min_{h\in\mathbb{R}} \Phi_\delta(\beta,h) - \frac{\log 2}{\beta} + \frac{\alpha}{2} + \frac{\alpha}{\beta} \log(\delta - 4\sqrt{\alpha}(1-\delta)),$$

and

$$\Psi^2(\beta,\alpha) = \min_{h\in\mathbb{R}} \Phi_0(\beta,h) - \frac{\log 2}{\beta} + \frac{\alpha}{2}.$$

Then, for every $\beta > 0$,

1. For $\mathbb{P}$-almost every $\omega \in \Omega$,

$$\Psi^1(\beta,\alpha,\delta) \leq \liminf_{\substack{N\to\infty \\ \frac{m}{N}\to\alpha}} f_{N,m,\beta} \leq \limsup_{\substack{N\to\infty \\ \frac{m}{N}\to\alpha}} f_{N,m,\beta} \leq \Psi^2(\beta,\alpha),$$

2.

$$\Psi^1(\beta,\alpha,\delta) \leq \liminf_{\substack{N\to\infty \\ \frac{m}{N}\to\alpha}} \mathbb{E}f_{N,m,\beta} \leq \limsup_{\substack{N\to\infty \\ \frac{m}{N}\to\alpha}} \mathbb{E}f_{N,m,\beta} \leq \Psi^2(\beta,\alpha),$$

In particular, when $\alpha = 0$, the free energy converges almost surely to a constant (the Curie-Weiss free energy) as it can be seen from the previous theorem by taking $\alpha = 0$ and letting, by continuity, $\delta\to 0$.

The weaker assertion (2) of the previous theorem — the majorisation and minoration of $\mathbb{E}f_{N,m,\beta}$ — and the idea of its proof were formulated in [49]. In the present form it was completely proved in [18, 103]. The complete proof is quite complicated and needs several intermediate steps for which we direct the interested reader to the original papers. Notice also that the existence of the quenched free energy, when $\alpha \neq 0$, remains an open problem in spite of the continuing efforts. This is a general problem not only for Hopfield model but also for the frustrated mean-field spin-glasses (see [82] for instance).

5.4. SELF-AVERAGING

Self-averaging is a property of some macroscopic quantities of disordered systems. Here we only stick to the self-averaging property of the quenched free energy.

It is remarkable that free energy of the Hopfield model has the *weak self-averaging property*; it is shown that the quenched free energy is very "close" to its average. Thus although the existence of the average it is not known, it can be proven that the free energy satisfies a very strong concentration property.

This result has a long history. Using martingale difference methods introduced by Girko for the study of spectral properties of random matrices in [35] (see [37] for a more recent and more easily accessible document), Pastur and Shcherbina obtain, in [77], a weak self-averaging property for the quenched free energy of the Sherrington-Kirkpatrick spin glass. Then, their method has been applied to the Hopfield model by Vermet, in [102], and then this result has been improved by Shcherbina and Tirozzi, in [93], and then in [78]. Following the same lines of reasoning but using a much more careful estimation of the terms appearing in the expansions, Bovier, Gayrard, and Picco [18] obtained the strongest formulation of weak self-averaging. Their theorem essentially proves the almost sure convergence of a rate function that is related to a constrained free energy. Stating precisely their result at this stage should need the introduction of various notions and notations that will be naturally introduced in the next chapter. Therefore, formulation of this result is postponed. Instead, it is preferred to state and prove here a weaker [102] self-averaging result that has the advantage of illustrating the Girko's method quite transparently.

Theorem 5.4. Let $\beta > 0$ and $m \leq \gamma(N)\sqrt{N}$, with $\lim_N \gamma(N) = 0$. Then,

$$\lim_{N\to\infty; m\to\infty} \mathbb{E}\left((f_{N,m,\beta} - \mathbb{E}f_{N,m,\beta})^2\right) = 0.$$

Proof. Introduce an auxiliary parameter $t \in [0,1]$ and fix some site k with $1 \leq k \leq N$. Instead of the original Hopfield Hamiltonian

$$H_{N,m}(x) = -\frac{1}{2N}\sum_{\mu=1}^{m}\sum_{i,j=1,\dots,N; i\neq j} \xi_i^\mu \xi_j^\mu x_i x_j,$$

consider the modified Hamiltonian

$$\widetilde{H}_{N,m}(x;k,t) = -\frac{1}{2N}\sum_{\mu=1}^{m}\sum_{i\neq j; i,j\neq k} \xi_i^\mu \xi_j^\mu x_i x_j - \frac{t}{N}\sum_{\mu=1}^{m}\sum_{i=1,\dots,N; i\neq k} \xi_i^\mu \xi_k^\mu x_i x_k.$$

Obviously, $\widetilde{H}_{N,m}(x;k,1) = H_{N,m}(x)$.

With this modified Hamiltonian, define the modified partition function

$$\widetilde{Z}_{N,m,\beta}(k,t) = \sum_x \exp(-\beta\widetilde{H}_{N,m}(x;k,t)),$$

the modified finite volume, quenched, specific, free energy

$$\widetilde{f}_{N,m,\beta}(k,t) = -\frac{1}{\beta N} \log \widetilde{Z}_{N,m,\beta}(k,t),$$

and the modified finite volume Gibbs measure

$$\widetilde{\gamma}_{N,m,\beta}(\cdot;k,t) = \frac{1}{\widetilde{Z}_{N,m,\beta}(k,t)} \sum_x \exp(-\beta \widetilde{H}_{N,m}(x;k,t)) \mathbb{1}_{(\cdot)}(x).$$

Following Girko, introduce now the increasing sequence of σ-algebras

$$\mathcal{F}_k = \sigma(\xi_1, \dots, \xi_k), \quad k = 0, \dots, N,$$

with $\mathcal{F}_0 = \{\emptyset, \Omega\}$. It is immediate that

$$\begin{aligned} f_{N,m,\beta} - \mathbb{E} f_{N,m,\beta} &= \sum_{k=1}^{N} [\mathbb{E}(f_{N,m,\beta}|\mathcal{F}_k) - \mathbb{E}(f_{N,m,\beta}|\mathcal{F}_{k-1})] \\ &= \sum_{k=1}^{N} [\mathbb{E}(\widetilde{f}_{N,m,\beta}(k,1)|\mathcal{F}_k) - \mathbb{E}(\widetilde{f}_{N,m,\beta}(k,1)|\mathcal{F}_{k-1})] \end{aligned}$$

because $f_{N,m,\beta}$ is measurable with respect to $\mathcal{F}_N$. Moreover, the terms of this sum are orthogonal for different k's. Now, use the trivial identity, valid for any $g \in C^1([0,1])$,

$$g(1) = g(0) + \int_0^1 \frac{dg}{dt}(t)dt.$$

Applying this identity in the summands above we remark that

$$\mathbb{E}(\widetilde{f}_{N,m,\beta}(k,0)|\mathcal{F}_k) - \mathbb{E}(\widetilde{f}_{N,m,\beta}(k,0)|\mathcal{F}_{k-1}) = 0$$

because the part of $\widetilde{f}_{N,m,\beta}(k,t)$ that is measurable with respect to $\mathcal{F}_k$ without being measurable with respect to $\mathcal{F}_{k-1}$, is precisely the term

$$-\frac{t}{N} \sum_{\mu=1}^{m} \sum_{i=1,\dots,N; i \neq k} \xi_i^\mu \xi_k^\mu x_i x_k$$

which vanishes when $t = 0$. It only remains the derivative part yielding

$$f_{N,m,\beta} - \mathbb{E}f_{N,m,\beta} =$$
$$\frac{1}{N^2}\sum_{k=1}^{N}\int_0^1\sum_{\mu=1}^{m}\sum_{i;i\neq k}[\mathbb{E}(\xi_i^\mu\xi_k^\mu\widetilde{\gamma}_{N,m,\beta}(x_ix_k;k,t)|\mathcal{F}_{k-1})-\xi_k^\mu\mathbb{E}(\xi_i^\mu\widetilde{\gamma}_{N,m,\beta}(x_ix_k;k,t)|\mathcal{F}_k)]dt.$$

Bounding trivially

$$|\widetilde{\gamma}_{N,m,\beta}(x_ix_k;k,t)| \leq 1,$$

we get from the orthogonality of the martingale differences that

$$\mathbb{E}\left((f_{N,m,\beta} - \mathbb{E}f_{N,m,\beta})^2\right) \leq \frac{4m^2}{N},$$

hence the theorem. ∎

Remark. It is worth noticing that to improve the above result, the martingale difference part of the proof remains unchanged. Only some technical additional work is needed to improve the trivial bound $|\widetilde{\gamma}_{N,m,\beta}(x_ix_k;k,t)| \leq 1$ used to estimate the summands. This can be achieved by using combinatorial and spectral estimates [93, 103].

At this place, a special mention is needed for the profound results of Talagrand [99], interested in precise estimates valid not only asymptotically but at every N. Based on techniques of concentration of measures [58, 97, 98], he establishes, among other things, a series of results on concentration properties of the free energy.

Finally in [90], using previously developed methods [2], it is shown that $Z_{N,m,\beta}/\mathbb{E}Z_{N,m,\beta}$ converges, in the high temperature regime, to a log-normal random variable when $m = \alpha N$, with $\alpha > 0$. Although I believe this result is undoubtedly correct, the proof given in [90], especially the proof of the technical result 2.17′ given in their appendix — on which the estimates of the whole paper rely —, is in my opinion incomplete. More recently, Talagrand [99] proves a stronger result, namely

Theorem 5.5. Let $\alpha = m/N$ and $\beta(1+\sqrt{\alpha}) < 1$. Then there is a positive constant K depending on α and β such that, for every $u > 0$,

$$\mathbb{P}\left(f_{N,m,\beta} \geq +\frac{m}{2\beta N}(\log(1-\beta) - u)\right) \leq K \exp(-\frac{u^2}{K}).$$

This theorem is valid at every N and not only asymptotically.

6. Gibbs States of Neural Networks

6.1. EXTENSION OF MEASURES

Extension of a measure defined on a finite-dimensional space to a measure on an infinite dimensional one is a basic problem in probability theory since the mere existence of stochastic processes lies on such a construction. The first result towards this direction was the celebrated Kolmogorov's extension theorem [51]. It is instructive to recall the precise statement of Kolmogorov's theorem and to introduce some notation that will be useful in the sequel.

6.1.1. *Fixing marginals: Kolmogorov's extension and unicity.* As usual, the one site state space is a probability space $(S, \mathcal{S}, \kappa)$ that will be assumed *compact* and metrisable. For a vertex set V (that is a denumerable discrete set and assumed embedable in a Lipschitz way into $\mathbb{R}^d$ for some d), the configuration space is given by $X = \{x : V \to S\} = S^V$. For a finite subset $\Lambda \subset V$, we denote $X_\Lambda = \{x : \Lambda \to S\} = S^\Lambda$; this space will be identified with the set of restrictions x_Λ of configurations on Λ. The natural σ-algebra on X_Λ will be $\mathcal{F}_\Lambda = \mathcal{S}^\Lambda$ and product measure will be denoted κ^Λ. We use the symbol $\mathcal{F}$ for $\mathcal{F}^V$. The problem of extension of measure is the possibility of equipping the measurable space $(X, \mathcal{F})$ with a probability $\mathbb{P}$ whose finite-dimensional marginals are fixed.

Theorem 6.1 (Kolmogorov). Let $(\Lambda_n)_{n\geq 1}$ be an increasing sequence of finite subsets of V and $(X_{\Lambda_n}, \mathcal{F}_{\Lambda_n})$ be the corresponding sequence of measurable configuration spaces. Suppose that on each $\mathcal{F}_{\Lambda_n}$ is defined a probability $\mathbb{P}_{\Lambda_n}$ such that the family $(\mathbb{P}_{\Lambda_n})_n$ verifies the following compatibility condition:

$$\text{If } \Lambda_n \subseteq \Lambda_{n'} \text{ then } \mathbb{P}_{\Lambda_n}(F) = \mathbb{P}_{\Lambda_{n'}}(F \times S^{\Lambda_{n'} \setminus \Lambda_n}), \quad \forall F \in \mathcal{F}_{\Lambda_n}$$

(the measure $\mathbb{P}_{\Lambda_n}$ is the marginal of $\mathbb{P}_{\Lambda_{n'}}$ on X_{Λ_n}). Then there exists a *unique* probability measure $\mathbb{P}$ on $(X, \mathcal{F})$ such that

$$\mathbb{P}(F \times S^{\Lambda_n^c}) = \mathbb{P}_{\Lambda_n}(F), \quad \forall F \in \mathcal{F}_{\Lambda_n}$$

(the finite-dimensional marginals of $\mathbb{P}$ coincide with $\mathbb{P}_{\Lambda_n}$).

This theorem, a corner-stone in the theory of stochastic processes, is of restricted use in the context of statistical mechanics because it excludes the possibility of phase transitions.

6.1.2. *Fixing conditional expectations: The DLR construction and phase transition.* The DLR construction, after the names of Dobrushin [26] and Lanford and Ruelle [56], is reminiscent of the Kolmogorov's construction. Instead of fixing finite-dimensional marginals however, we fix conditional probabilities with respect to fixed boundary conditions. The DLR construction can be seen as a generalisation of the existence of an invariant probability for a Markov chain. Here however, the ordering of the indexing set, played by "time" for Markov chains, is played by an ordering by inclusions of subsets of finite subsets of V. The Markovian character of the process is guaranteed by the existence of a genuine Hamiltonian as it is explained below.

Notations 6.2. We use the symbol $A \subset\subset V$ to denote that A is a *finite* subset of V.

Definition 6.3. Let $(\Phi_A)_{A \subset\subset V}$ be a family of mappings $\Phi_A : X \to \mathbb{R}$, indexed by the finite subsets of V such that Φ_A is $\mathcal{F}_A$-measurable for every $A \subset\subset V$ (the mapping Φ_A depends only on the configurations restricted over A). Then the family $(\Phi_A)_{A \subset\subset V}$ is called a family of *interaction potential.*

Interaction potentials introduce a coupling between configurations over different sites that allow to construct measures that are not plainly product measures. However, to keep an overall spatial Markovian structure, this interaction must be moderate.

Definition 6.4. If, for every $\Lambda \subset\subset V$ and every $x \in X$, the sum

$$\sum_{A \subset\subset V; A \cap \Lambda \neq \emptyset} \Phi_A(x)$$

exists, then it is called (genuine) *interaction Hamiltonian* and is denoted by $H_\Lambda(x)$. If this sum does not exist but there is an increasing and diverging to $+\infty$ sequence of positive numbers $(a_\Lambda)_{\Lambda \subset\subset V}$ such that

$$\frac{1}{a_\Lambda} \sum_{A \subset \Lambda} \Phi_A(x)$$

is extensive, then the latter sum is called *mean-field interaction Hamiltonian* and is denoted by the same symbol $H_\Lambda(x)$.

Remark. In the above definition, *extensive* means that, asymptotically and in some sense not to be farther precised here, the mean-field Hamiltonian $H_\Lambda(x)$ behaves like $|\Lambda|$.

Example 6.5. Let $S = \{-1, 1\}$. For a positive parameter J and a real parameter h, we define the interaction potentials

$$\Phi_A(x) = \begin{cases} -Jx_i x_j & \text{if } A = \{i,j\} \text{ and } |i-j| = 1 \\ -hx_i & \text{if } A = \{i\} \\ 0 & \text{otherwise} \end{cases}$$

and

$$\Psi_A(x) = \begin{cases} -Jx_i x_j & \text{if } A = \{i,j\} \\ -hx_i & \text{if } A = \{i\} \\ 0 & \text{otherwise.} \end{cases}$$

The potential Φ gives rise to a genuine Hamiltonian and defines the ferromagnetic Ising model with external field; the potential Ψ gives rise to mean-field Hamiltonian with $a_\Lambda = |\Lambda|$ and defines the Curie-Weiss model with external field.

We assume henceforth in this subsection that the model is defined by a genuine interaction Hamiltonian and that for every $\beta > 0$ and every y_{Λ^c}, the integral

$$Z_{\Lambda^c,\beta}(y_{\Lambda^c}) = \int_{X_\Lambda} \exp\left(-\beta H_\Lambda(x_\Lambda y_{\Lambda^c})\right) \kappa^\Lambda(dx)$$

exists. Here $x_\Lambda y_{\Lambda^c}$ denotes the concatenation of two restricted configurations, that is a configuration $z \in X$ such that

$$z_i = \begin{cases} x_i & \text{if } i \in \Lambda \\ y_i & \text{if } i \notin \Lambda \end{cases}$$

The configuration y_{Λ^c} is a boundary condition and the quantity $Z_{\Lambda^c,\beta}(y_{\Lambda^c})$ is called the *finite-volume partition function* with boundary condition y_{Λ^c}.

Definition 6.6. A probability defined on $(X, \mathcal{F})$ for every measurable set $F \in \mathcal{F}$ and any boundary condition $y_{\Lambda^c} \in X_{\Lambda^c}$ by

$$\gamma_{\Lambda,\beta}(F|y_{\Lambda^c}) = \frac{1}{Z_{\Lambda^c,\beta}(y_{\Lambda^c})} \int_{X_\Lambda} \exp\left(-\beta H_\Lambda(x_\Lambda y_{\Lambda^c})\right) \mathbb{1}_F(x_\Lambda y_{\Lambda^c}) \kappa^\Lambda(dx)$$

is called a *finite-dimensional Gibbs' specification* for the boundary condition $y_{\Lambda^c} \in X_{\Lambda^c}$.

Remark. The Gibbs' specification is a Markovian kernel of conditional probabilities with respect to the ordering defined by the inclusions.

We are seeking for measures γ on $(X, \mathcal{F})$ having as finite-dimensional conditional laws the Gibbs' specifications.

Definition 6.7 (DLR equation). Let

$$\mathcal{G} = \{\gamma \in \mathcal{M}_1(X, \mathcal{F}) : \forall F \in \mathcal{F}, \gamma\text{-a.e. } y \in X, \forall \Lambda \subset\subset V; \gamma(F|y_{\Lambda^c}) = \gamma_{\Lambda,\beta}(F|y_{\Lambda^c})\}.$$

The set $\mathcal{G}$ is called set of *Gibbs measures* specified by the family of kernels $(\gamma_{\Lambda,\beta}(\cdot|y_{\Lambda^c}))_{\Lambda \subset\subset V}$.

Remark. Contrary to the Kolmogorov's theorem, the DLR equation is less restrictive: the set $\mathcal{G}$ can be empty, can be a singleton, or it can have several elements (in fact infinitely many). For models encountered in statistical mechanics for which the DLR construction is possible, the set $\mathcal{G}$ is not empty. The passage from the regime of unique Gibbs measure to the regime where many solutions exist is called a *phase transition*.

The general structure of the set $\mathcal{G}$ is a difficult problem and is the object of study of a whole discipline, the equilibrium statistical mechanics, lying beyond the scope of the present review. The interested reader can profitably consult complete treatises on the topic, [34, 87, 94]. The only thing that will be mentioned here is that the set $\mathcal{G}$ has a simplicial structure. The extremal points of this set, called *pure states*, are attained through special choices of the boundary conditions; in general, the limiting Gibbs measure is a convex combination of pure states.

We end however this subsection by recalling once more that the DLR construction is possible only for models defined by a genuine interaction Hamiltonian excluding all mean-field neural network models. For instance, the only neural network model from the ones presented in this report for which DLR construction is possible is the one called "short-range finite-dimensional network", introduced in section 2.6.

6.1.3. *Weak limiting procedure.* For models defined by genuine Hamiltonians, still another construction is possible that is shown to be equivalent to the DLR construction. This is the *weak-limiting* procedure, defined briefly below.

Consider a system with genuine finite-volume Hamiltonian H_Λ. Fix some arbitrary configuration $y \in X$ and define the relative Hamiltonian $H_\Lambda(\cdot|y) : X \to \mathbb{R}$ by

$$H_\Lambda(x|y) = H_\Lambda(x_\Lambda y_{\Lambda^c}).$$

Denote by $\sigma_\Lambda : X \to X_\Lambda$ the *canonical projection* such that $x \mapsto \sigma_\Lambda(x) = x_\Lambda \in X_\Lambda$. Every one site space S (viewed as a fibre for the construction of the configuration space as a fibre bundle over the base V), can be equipped with the discrete topology τ_0. Thus, X can be equipped with the product topology $\tau = \prod_{i\in V} \tau_i$; with this topology, the canonical projections are continuous functions.

Now the Borel σ-algebra $\mathcal{B}(X)$, rendering measurable the open sets for the topology τ of X, coincide with the σ-algebra $\mathcal{F}$ generated by the collection $\mathcal{C}$ of *cylinder sets* $C_{\Lambda_n}(F_n)$, where for a given $n \in \mathbb{N}$ and a given increasing sequence of volumes Λ_n, we define the cylinder sets by

$$C_{\Lambda_n}(F_n) = \{x \in X : \sigma_{\Lambda_n}(x) \in F_n \quad \text{for} \quad F_n \in \mathcal{F}_{\Lambda_n}\}.$$

Define also, for a fixed configuration $y \in X$, the space

$$X_\Lambda(y) = \{x \in X : x_i = y_i \text{ for } i \in \Lambda^c\}.$$

With the definitions introduced so far, it is possible to introduce a sequence of probability measures on $(X, \mathcal{B}(X))$ by

$$\mu_{n,\beta,y}(A) = \gamma_{\Lambda_n,\beta,y_{\Lambda_n^c}}(\sigma_{\Lambda_n}(A \cap X_{\Lambda_n}(y))), \quad \forall A \in \mathcal{B}(X),$$

where $\gamma_{\Lambda_n,\beta,y_{\Lambda_n^c}}$ is a finite-volume Gibbs measure defined by

$$\gamma_{\Lambda,\beta,y_{\Lambda^c}}(x_\Lambda) = \frac{\exp(-\beta H_\Lambda(x_\Lambda|y_{\Lambda^c})}{\sum_{x_\Lambda \in X_\Lambda} \exp(-\beta H_\Lambda(x_\Lambda|y_{\Lambda^c})}.$$

Notice that $\gamma_{\Lambda_n,\beta,y_{\Lambda_n^c}}$ is very reminiscent of the Gibbs specification introduced in the previous section but it does not exactly coincides with it since it is not defined on the whole configuration space X but only the subspace X_Λ.

The space S being compact for the discrete topology, the same holds true, by virtue of the Tychonov's theorem, for the space X for the product topology τ. Moreover, the topology τ can be easily metrised so that X is a compact metrisable space, hence complete and separable, what technically is called a *Polish space*. By Riesz-Markov theorem, there exists a bijection between probability measures μ on $(X, \mathcal{B}(X))$ and positive normalisable linear functionals on the Banach space $C(X)$ of continuous bounded real functions on X. Thus the set $\mathcal{M}_1(X)$ of probability measures on X is identified with a subset of the dual $C^*(X)$ of $C(X)$. Now the topology τ defines continuity on $C(X)$; it induces therefore a weak-* on $C^*(X)$, called the *vague topology* on $\mathcal{M}_1(X)$. (Recall [12] that a sequence of probability measures (μ_n) of $\mathcal{M}_1(X)$ converges weakly to μ, denoted $\mu_n \Rightarrow \mu$ if, and only if, $\lim_{n\to\infty} \int f d\mu_n \to \int f d\mu$, for all $f \in C(X)$.) We have the following

Theorem 6.8. The space $\mathcal{M}_1(X)$ is compact for the topology of weak convergence. Moreover, $\mu_n \Rightarrow \mu$ if, and only if, $\mu_n(C) \to \mu(C)$ for every cylinder set C of X.

The previous theorem guarantees that all sequences of finite volume measures have at least one accumulation point. We thus define

Definition 6.9. A probability μ on $(X, \mathcal{B}(X))$ is a Gibbs measure if μ belongs to the closed convex envelope of the set of accumulation points (for the weak topology) of the sequence of measures $(\mu_{n,\beta,y}(\cdot))_n$ defined above.

This construction gives rise to Gibbs measures that are convex combinations of extremal (pure) DLR Gibbs states [67], defined through fixed boundary conditions. The natural question is how to choose *extremal* states through this weak limiting procedure. This can be achieved by perturbing the original Hamiltonian by additional terms that vanish eventually, after the infinite-volume limit is taken. To fix ideas, consider the standard Ising model Hamiltonian

$$H_\Lambda(x) = -J \sum_{i\in\Lambda} \sum_{j\in\mathbb{Z}^d; |i-j|=1} x_i x_j .$$

This Hamiltonian has two minima[6], denoted respectively $+$ and $-$, defined so that

$$x = + \Leftrightarrow x_i = +1, \forall i \in \Lambda$$
$$x = - \Leftrightarrow x_i = -1, \forall i \in \Lambda$$

The well known Peierls argument establishes the existence of a phase transition (existence of at least two different infinite volume Gibbs measures) provided that $d \geq 2$. Consider now a Hamiltonian perturbed by an additional term

$$H_\Lambda^h(x) = -J \sum_{i\in\Lambda} \sum_{j\in\mathbb{Z}^d; |i-j|=1} x_i x_j - h \sum_{i\in\Lambda} x_i .$$

It is clear that if $h \neq 0$, one of the two ground energy configurations is favoured and thus the external field lifts the degeneracy. As a result, H_Λ^h, with $h \neq 0$ gives rise to a *unique* Gibbs measure. Taking now the limit $h \downarrow 0$ *after the infinite volume limit is performed*, the extremal Gibbs measure corresponding to the $+$ boundary condition DLR Gibbs state is obtained. Similarly by taking $h \uparrow 0$, the DLR Gibbs state with $-$ boundary condition is chosen.

[6] This is true *stricto sensu* only in the infinite-volume limit. Now, in the infinite-volume limit the Hamiltonian diverges. The precise statement is that both $H_\Lambda(+)$ and $H_\Lambda(-)$ are of the order $-J|\Lambda|(1 + \mathcal{O}(\frac{|\partial\Lambda|}{|\Lambda|}))$

This observation is quite general and allows to choose extremal Gibbs measures. When there are q minimising configurations, $q-1$ external fields are needed to lift the degeneracy.

6.2. INDUCED MEASURES

Even the weak limiting procedure introduced in the last subsection needs the description of the model by a genuine Hamiltonian and thus fails for mean-field models. One method to circumvent this difficulty is to use null boundary conditions for mean-field models as it was done in [4] for the Curie-Weiss random field model. Another possibility is to reduce the problem to the equivalent problem over magnetisation and follow a precise weak limiting procedure introduced in [18] we describe briefly below.

We have already seen that $\gamma_{N,m,\beta}$, the finite volume Gibbs measure, is the unique invariant probability of stochastic evolution. However, the model is a long range system and this destroys its spatial Markovian properties and makes impossible the DLR construction.

Starting from the (slightly modified) Hopfield Hamiltonian

$$H_{N,m}(x) = -\frac{1}{2N}\sum_{i,j}\sum_{\mu=1}^{m}\xi_i^{\mu}\xi_j^{\mu}x_i x_j,$$

we introduce, for every $\eta \in \mathbb{N}$, the finite volume Gibbs measure in an external field h by

$$\gamma_{N,m,\beta}^{\eta,h}(x) = \frac{1}{Z_{N,m,\beta}^{\eta,h}}\exp\left(-\beta H_{N,m}(x) + \beta h\sum_{i}\xi_i^{\eta}x_i\right),$$

where $Z_{N,m,\beta}^{\eta,h}$ is the new normalising factor.

The Hamiltonian can also be expressed in terms of overlap parameters

$$v_N^{\mu}(x) = \frac{1}{N}\sum_{i}\xi_i^{\mu}x_i, \quad \text{for} \quad \mu = 1,\dots,m,$$

by writing

$$H_{N,m}(x) = -\frac{N}{2}\sum_{\mu=1}^{m}\left(v_N^{\mu}(x)\right)^2 = -\frac{N}{2}\left\|v_N(x)\right\|_2^2.$$

Therefore, the Hamiltonian depends on the configurations x only through the overlap parameters v. It is convenient to introduce *induced measures* on $\mathbb{R}^m$, instead of ordinary Gibbs measures, by

$$g_{N,m,\beta}^{\eta,h}(v) = \gamma_{N,m,\beta}^{\eta,h}(\{v(x) = v\})$$

for $v = (v^1, \ldots, v^m)$. For $\delta > 0$, write $a(\delta, \beta)$ for the largest solution of the equation $\delta a = \tanh(\beta a)$ and denote by $\|\cdot\|$ the ℓ^2 norm on $\mathbb{R}^{\mathbb{N}}$. Let $\lim \frac{m}{N} = \alpha$. For fixed β, for $\nu \in \mathbb{N}$, and $S \in \{-1, 1\}$, we define the ball

$$B_\rho^{(\nu,s)} = \{x \in \mathbb{R}^{\mathbb{N}} : \|x - sa(1 - 2\sqrt{\alpha}, \beta)e^\nu\| \leq \rho\},$$

where e^ν denotes the ν-th unit vector in $\mathbb{R}^{\mathbb{N}}$. With this notation, it is proven in [18] the following

Theorem 6.10. There exists $\alpha_0 > 0$ such that, for every $\alpha \leq \alpha_0$ and every $\beta > 1 + 3\sqrt{\alpha}$, if $\rho^2 > C[a(1 - 2\sqrt{\alpha}, \beta)]^{3/2}\alpha^{1/8}|\log \alpha|^{1/4}$, almost surely,

$$\lim_{h\downarrow 0} \lim_{N\uparrow\infty} g_{N,m,\beta}^{\eta,h}(B_\rho^{(\eta,+1)}) = 1.$$

So, this theorem guarantees that by using a small external field, eventually vanishing to zero, we can force the induced measure to be concentrated on a ball of overlap parameters slightly differing from a selected direction η. This result is the first rigorous indication that stochastic dynamics can be used to recover associatively the stored patterns in a large network. However, important problems remain open concerning the dynamical evolution of the net. For instance, it is not yet rigorously established whether a simulated annealing algorithm can be used to explore the stored patterns. Although such a possibility is expected, it remains to know how the various parameters have to be adjusted for such an algorithm to converge.

We are now able to state the result of [18] concerning the weak self-averaging property of the rate function. Denote

$$\phi_{N,\beta,\rho}(\widetilde{v}) = -\frac{1}{\beta N} \log g_{N,\beta,h=0}(\|v_\Lambda - \widetilde{v}\|_2^2 \leq \rho).$$

The function $\phi_{N,\beta,\rho}$ is the large deviation rate function governing the exponential convergence to zero of the corresponding probability. It is closely related to the free-energy since the latter can be expressed in a similar way where the restricting event $\{\|v_\Lambda - \widetilde{v}\|_2^2 \leq \rho\}$ is replace by the whole space.

It is proven in [18] the following

Theorem 6.11. Assume that $\lim_{N\to\infty} \frac{m}{N} = \alpha > 0$. Let $\rho < 1$ and $\|\widetilde{v}\|_2$ be bounded. Then, for every $n \in \mathbb{N}$, there exists $t_n < \infty$ such that $\forall t \geq t_n$, and for N large enough,

$$\mathbb{P}\left(|\phi_{N,\beta,\rho}(\widetilde{v}) - \mathbb{E}\phi_{N,\beta,\rho}(\widetilde{v})| \geq \frac{t(\log N)^{3/2}}{\sqrt{N}} \right) \leq \frac{1}{N^n}.$$

The proof of this theorem consists in writing $\phi_{N,\beta,\rho}(\widetilde{v}) - \mathbb{E}\phi_{N,\beta,\rho}(\widetilde{v})$ as a martingale difference, following the ideas of Girko explained in the previous chapter, and then use precise estimates for these martingales differences. The proof is somewhat technical to be reproduced *in extenso* here and the interested reader is directed to the original publication.

7. Conclusion

In this review, the profound relation existing between neural networks and statistical mechanics is shown. Due to the limited space and time available, only a selection of mathematical results is presented here.

Neural nets have found an extended field of applications in constructing engineering devices where they are used as an alternative to universal computers to perform tasks of categorisation, pattern recognition, forecasting, and so on.

All these fascinating applications are missing from the present report. One may consult the books [53, 70, 45] to get a flavour of possible applications and the specialised journals like *Network, IEEE Neural Networks, IEEE Patterns Analysis and Machine Intelligence, ...* for some more finalised applications.

Biologists and neurophysiologists also use neural nets modelling to explain the functioning of the brain. Experiments in neurophysiology identify new characteristics of neural behaviour and these are incorporated into more and more sophisticated neural models. All these results are missing from this review.

From the moment that the connection of neural networks with statistical mechanics was established, the subject became a branch of theoretical physics. Many interesting and intuitively appealing results were obtained by the physical community. Based on heuristic approaches like the replica trick, many qualitatively convincing results have been obtained. All these results are also missing from this survey. The interested reader is directed to [66, 25, 32, 31, 69, 54] for the most important of them.

The reader could expect therefore that all the mathematical aspects would be presented here. She will be disappointed: only results establishing that learning process is equivalent to an information increasing (entropy decreasing) process and that the restitution process is a Markov process converging to an invariant measure interpreted as the Gibbs measure of statistical mechanics are presented. Interesting issues as those describing dynamical evolution of neural nets with stochastic differential equations [10, 47] or with discrete time evolutions [11, 21] are omitted. But also mathematical results connected to more specific issues of neural networks such as biomathematical modelling of neural functions [24] or connections with graph theory are absent from this report.

The main reason for these omissions is that the subject is so vast that without limiting oneself in some clearly circumscribed region their is a risk of endless ramification. The second reason is that even in the restricted domain of thermodynamic formalism examined here, there are still numerous open problems. The known mathematical results only partially explain the numerically observed phenomena. Many questions pointed out in the main text — like the storage capacity, the stability and convergence under several steps of the dynamics, the study of systems with dependent patterns, etc. — remain unanswered for the moment. The mathematical methods presented here offer may be a good starting point to tackle these questions but new bright ideas are also certainly needed.

Acknowledgements

The author wishes to thank the Departamento de Ingeniería Matemática de la Universidad de Chile for its kind invitation to give this series of lectures. He also acknowledges support from the EU network CHRX-CT93-0411 that produced a significant amount of the information contained here and favoured its dissemination. He wishes also to express his acknowledgements to Anton Bovier, Christof Külske, and Franck Vermet for their careful reading of the manuscript and their comments.

References

[1] Abbott, L.F., Learning in Neural Network Memories, *Network* **1**, 105–122 (1990).

[2] Aizenman, M., J.L. Lebowitz, D. Ruelle, Some Rigorous Results on the Sherrington-Kirkpatrick Model, *Commun. Math. Phys.*, **112**, 3–20 (1987).

[3] Albeverio, S., B. Tirozzi, B. Zegarlinski, Rigorous Results for the Free Energy in the Hopfield Model, *Commun. Math. Phys.*, **150**, 337–373 (1992).

[4] Amaro de Matos, J.M.G., A.E. Patrick, V.A. Zagrebnov, Random Infinite Volume Gibbs States for the Curie-Weiss Random Field Ising Model, *J. Stat. Phys.*, **66**, 139–164 (1992).

[5] Amit, D.J., *Modelling Brain Function*, Cambridge University Press, Cambridge (1989).

[6] Amit, D.J., G. Gutfreund, H. Sompolinsky, Statistical Mechanisms of Neural Networks near Saturation, *Ann. Phys.*, **173**, 30–67 (1987).

[7] Anily, S., A. Federgruen, Simulated Annealing Methods with General Acceptance Probabilities, *J. Appl. Prob.* **24**, 657–667 (1968).

[8] Aubin, J.P., *Mathematical Methods for Neural Networks*, lecture notes of a COMETT graduate module held in Les Houches, 16–29 March (1992).

[9] Azencott, R., Simulated Annealing, *Séminaire Bourbaki No. 697*, 1–15 (1988).

[10] Ben Arous, G., A. Guionnet, *Large Deviations for Langevin Spin Glass Dynamics*, preprint DMI, École Normale Supérieure (1995).

[11] Bernier, O., Stochastic Analysis of the Dynamics of a General Class of Synchronous Neural Networks, *J. Phys. A: Math. Gen.*, **26**, 6879–6892 (1993).

[12] Billingsley, P., *Convergence of Probability Measures*, Wiley, New York (1968).

[13] Bovier, A., Self-Averaging in a Class of Generalised Hopfield Models, *J. Phys. A: Math. Gen.* **27**, 7069–7077 (1994).

[14] Bovier, A., V. Gayrard, Rigorous Bounds on the Storage Capacity of the Dilute Hopfield Model, *J. Stat. Phys.* **69**, 597–627 (1992).

[15] Bovier, A., V. Gayrard, Rigorous Results on the Thermodynamics of the Dilute Hopfield Model, *J. Stat. Phys.*, **72**, 79–112 (1993).

[16] Bovier, A., V. Gayrard, *An almost sure Large Deviation Principle for the Hopfield Model*, Weierstraß IAAS preprint (1995).

[17] Bovier, A., V. Gayrard, P. Picco, Gibbs States of the Hopfield Model in the Regime of Perfect Memory, *Prob. Th. Rel. Fields*, **100**, 329–363 (1994).

[18] Bovier, A., V. Gayrard, P. Picco, Gibbs States of the Hopfield Model with Extensively Many Patterns, *J. Stat. Phys.*, **79**, 395–414 (1995).

[19] Bovier, A., V. Gayrard, P. Picco, *Large Deviation Principles for the Hopfield and the Kac-Hopfield model*, Weierstraß IAAS preprint (1994).

[20] Broeck, C. van den, Statistical Physics of Learning from Examples: a Brief Introduction, *Acta Phys. Polon. B*, **25**, 903–923 (1994).

[21] Cassandro, M., A. Galves, E. Olivieri, M. E. Vares, Metastable Behaviour of Stochastic Dynamics: a Pathwise Approach, *J. Stat. Phys.*, **35**, 603– (1984).

[22] Changeux, J.-P., *L'Homme Neuronal*, Fayard, Paris (1983).

[23] Collet, P., J.-P. Eckmann, *Iterated Maps on the Interval as Dynamical Systems*, Birkhäuser, Basel (1980).

[24] Cottrell, M., Mathematical Analysis of a Neural Network with Inhibitory Coupling, *Stoch. Proc. Appl.*, **40**, 103–126 (1992).

[25] Derrida, B., R.B. Griffiths, A Prügel-Bennett, Finite-Size Effects and Bounds for Perceptron Models, *J. Phys. A: Math. Gen.*, **24**, 4907–4940 (1991).

[26] Dobrushin, R.L., The Description of a Random Field by Means of Conditional Probabilities and Condition of its Regularities, *Th. Prob. Appl.*, **13**, 458–486 (1968).

[27] Fassnacht, C., A. Zippelius, A recognition and categorisation in a structured neural network with attractor dynamics, *Network*, **2**, 63–84 (1992).

[28] Ferrari, P.A., S. Martínez, P. Picco, A Lower Bound for the Memory Capacity in the Potts-Hopfield Model, *J. Stat. Phys.*, **66**, 1643–1652 (1992).

[29] Folk, R., A. Kartashov, P. Lisoněk, P. Paule, Symmetries in Neural Networks: a Linear Group Action Approach, *J. Phys. A: Math. Gen.*, **26**, 3159–3164 (1993).

[30] Freidlin, M.I., A.D. Wentzell, *Random Perturbations of Dynamical Systems*, Springer-Verlag, Berlin (1984).

[31] Gardner, E., The Space of Interactions in Neural Network Models, *J. Phys. A: Math. Gcn.*, **21**, 257–270 (1987).

[32] Gardner, E., B. Derrida, Optimal Storage Properties of Neural Network Models, *J. Phys. A: Math. Gen.*, **21**, 271–284 (1988).

[33] Gayrard, V., The Thermodynamic Limit of the Potts-Hopfield Model for Infinitely Many Patterns, *J. Stat. Phys.*, **68**, 977–1011 (1992).

[34] Georgii, H.-O., *Gibbs Measures and Phase Transitions*, Walter de Gruyter, Berlin (1988).

[35] Girko, V.L., *Random Matrices*, Vishcha Shkola, Izdat. Kiev Univ., Kiev (1975).

[36] Girko, V.L., Limit Theorems for Maximal and Minimal Eignevalues of Random Matrices, *Th. Prob. Appli.*, **35**, 680–695 (1988).

[37] Girko, V.L., *Theory of Random Determinants*, Kluwer, Dodrecht (1990).

[38] Goles, E., S. Martínez, *Neural and automata networks*, Kluwer Academic Publ., Dodrecht (1990).

[39] Goles, E., S. Martínez, *Statistical Physics, Automata Networks, and Dynamical Systems*, Kluwer Academic Publ., Dodrecht (1992).

[40] Griniasty, M., M.V. Tsodyks, D.J. Amit, *Conversion of Temporal Correlations between Stimuli to Spatial Correlations between Attractors*, preprint Università di Roma 1 (1992).

[41] Haario, H., E. Saksman, Simulated Annealing Process in General State Space, *Adv. Appl. Prob.*, **23**, 866–893 (1991).

[42] Hebb, D., *The Organisation of Behaviour: a Neurolophysiological Theory*, Wiley, New York (1949).

[43] Hemmen, J.L. van, D. Grensing, A. Huber, R. Kühn, Nonlinear Neural Networks I: General Theory, *J. Stat. Phys.*, **50**, 231–257 (1988).

[44] Hemmen, J.L. van, D. Grensing, A. Huber, R. Kühn, Nonlinear neural networks II: information processing, *J. Stat. Phys.*, **50**, 259–293 (1988).

[45] Hertz, J., A. Krogh, R. Palmer, *Introduction to the theory of neural computation,* Addison-Wesley, Redwood City CA (1991).

[46] Hopfield, J.J., Neural Networks and Physical Systems with Emergent Collective Computational Abilities, *Proc. Natl. Acad. Sci.*, **79**, 2554–2558 (1982).

[47] Kallianpur, G., *Weak Convergence of Stochastic Neuronal Models*, in Stochastic Methods in Biology, Nagoya (1985), M. Kimura, G. Kallianpur, and T. Hida (eds.).

[48] Khinchin, A.I., *Mathematical Foundations of Information Theory*, Dover, New York (1957).

[49] Koch, H., A Free Energy Bound for the Hopfield Model, *J. Phys. A: Math. Gen.*, **26**, L353–L355 (1993).

[50] Koch, H., J. Piasko, Some Rigorous Results on the Hopfield Neural Network Model, *J. Stat. Phys.*, **55**, 903–928 (1989).

[51] Kolmogorov, A.N., *Foundations of Probability Theory*, Chelsea Publishing, New York (1950).

[52] Komlós, J., R. Paturi, Convergence Results in a Autoassociative Memory Model, *Neural Networks*, **1**, 239–250 (1988).

[53] Kosko, B., *Neural Networks and Fuzzy Systems*, Prentice-Hall, London (1992).

[54] Krauth, W., M. Mézard, Storage Capacity of Memory Networks with Binary Couplings, *J. Physique*, **50**, 3057–3066 (1989).

[55] Kuramoto, Y., *Chemical Oscillation, Waves, and Turbulence*, Springer-Verlag, Berlin (1984).

[56] Lanford III, O.E., D. Ruelle, Observables at Infinity and States with Short Range Correlations in Statistical Mechanics, *Commun. Math. Phys.* **13**, 194-215 (1969).

[57] Lang, S., *Real and Functional Analysis*, Springer-Verlag, Berlin (1993).

[58] Ledoux, M., M. Talagrand, *Probability in Banach Spaces*, Springer-Verlag, Berlin (1991).

[59] Little, W., G. Shaw, Analytic Study of the Memory Storage Capacity of a Neural Network, *Math. Biosc.*, **39**, 281–290 (1978).

[60] Loukianova, D., Capacité de Mémoire dans le Modèle de Hopfield, *C. R. Acad. Sci. Paris*, **318**, 157–160 (1994).

[61] Loukianova, D., *Étude Rigoureuse du Modèle de Mémoire Associative*, PhD Thesis, Université de Paris VII, presented on 2 December (1994).

[62] MacEliece, R., E. Posner, E. Rodemich, S. Venkatesh, The Capacity of the Hopfield Associative Memory, *IEEE Trans. Inf. Theory*, **33**, 461–482 (1987).

[63] Maes, C., V. Zagrebnov, On the Parallel Dynamics of a Multi-Layered Perceptron, (1991).

[64] Martínez, S., *Introduction to Neural Networks: Storage Capacity and Optimisation*, proceedings CIMPA school on "Dynamical and disordered systems", Temuco (1992).

[65] McCulloch, W.S., W. Pitts, A Logical Calculus of the Ideas Immanent in Nervous Activity, *Bull. Math. Biophys.* **5**, 115–133 (1943).

[66] Mézard, M., G. Parisi, M.A. Virasoro, *Spin-glass Theory and Beyond*, World scientific, Singapore (1988).

[67] Minlos, R., Gibbs' Limit Distribution, *Funct. Anal. Appl.*, **2**, 60–73; **3**, 40–53 (1967).

[68] Miyashita, Y., Neuronal Correlate of Visual Associative Long-Term Memory in the Primate Temporal Cortex, *Nature*, **335**, 817–819 (1988).

[69] Monasson, R., Properties of Neural Network Storing Spatially Correlated Patterns, *J. Phys. A: Math. Gen.*, **25**, 3701–3720 (1992).

[70] Müller, B., J. Reinhardt, *Neural Networks*, Springer-Verlag, Berlin (1990).

[71] Newman, C., Memory Capacity in Neural Networks, *Neural Networks*, **1**, 223–238 (1988).

[72] Nilson, N.J., *Learning Machines*, McGraw-Hill, New York (1965).

[73] Nishimori, H., W. Whyte, D. Sherrington, *Finite-Dimensional Neural Networks Storing Structured Patterns*, preprint University of Oxford (1994).

[74] Parisi, G., *Attractor Neural Networks*, preprint (1994) available from `cond-mat@babbage.sissa.it` under reference 9412030.

[75] Pastur, L.A., A.L. Figotin, Exactly Solvable Model of a Spin Glass, *Sov. J. Low Temp. Phys.*, **3**, 378–383 (1977).

[76] Pastur, L.A., A.L. Figotin, On the Theory of Disordered Spin Systems, *Theor. Math. Phys.*, **35**, 404–414 (1978).

[77] Pastur, L.A., M. Shcherbina, Absence of Self-Averaging of the Order Parameter in the Sherrington-Kirkpatrick Model, *J. Stat. Phys.*, **62**, 1–19 (1991).

[78] Pastur, L.A., M. Shcherbina, B. Tirozzi, The Replica Symmetric Solution without Replica Trick for the Hopfield Model, *J. Stat. Phys.*, **74**, 1161–1183 (1994).

[79] Patrick, A.E., V.A. Zagrebnov, Parallel Dynamics for an Extremely Diluted Neural Network, *J. Phys. A: Math. Gen.*, **23**, L1323–L1329 (1990).

[80] Patrick, A.E., V.A. Zagrebnov, A Probabilistic Approach to Parallel Dynamics for the Little Hopfield Model, *J. Phys. A: Math. Gen.*, **24**, 3413–3426 (1991).

[81] Petritis, D., *Simulations Numériques Monte Carlo*, preprint Université de Rennes I, to be published by Masson, Paris (1995).

[82] Petritis, D., *Equilibrium Statistical Mechanics of Frustrated Disordered Systems: a Survey of Mathematical Results*, preprint Université de Rennes I (1994), available from `cond-mat@babbage.sissa.it` under reference 9411098.

[83] Picco, P., *Artificial Neural Networks*, preprint CNRS Marseille (1995).

[84] Radons, G., H.G. Schuster, D. Werner, Fractal Measures and Diffusion as Results of Learning in Neural Networks, *Phys. Lett. A*, **174**, 293–297 (1993).

[85] Rockafellar, R.T., *Convex Analysis*, Princeton Univ. Press, Princeton (1970).

[86] Rosenblatt, F., *Principles of Neurodynamics*, Spartan, New York (1962).

[87] Ruelle, D., *Thermodynamic formalism*, Addison-Wesley, Reading (1978).

[88] Rumelhart, D.E., G.E. Hinton, R.J. Williams, Learning Representations by Back-Propagating Errors, *Nature*, **323**, 533– (1986).

[89] Sakaguchi, H., Learning Rules for an Oscillator Network, *Phys. Lett. A*, **174**, 289–292 (1993).

[90] Scacciatelli, E., B. Tirrozi, Fluctuation of the Free Energy in the Hopfield Model, *J. Stat. Phys.*, **67**, 981–1008 (1992).

[91] Scharze, H., J. Hertz, Learning from Examples in fully Connected Committee Machines, *J. Phys. A: Math. Gen.*, **26**, 4919–4936 (1993).

[92] Schlüter, M., E. Wagner, *Phys. Rev.*, **E49**, 1690– (1994).

[93] Shcherbina, M., B. Tirozzi, The Free Energy for for a Class of Hopfield Models, *J. Stat. Phys.*, **72**, 113–125 (1993).

[94] Simon, B., *The Statistical Mechanics of Lattice Gases*, Princeton University press, Princeton (1993).

[95] Solla, S., *Learning and Generalisation in Layered Neural Networks*, in Redes neuronales, teoría y aplicaciones, Escuela de Verano 88 en Física estadística y sistemas cooperativos (1988).

[96] Solla, S., A Theory of Supervised Learning, in *Neural networks: from biology to high energy physics*, Proc. Elba Int. Physics Centre, O. Benhar, C. Bosio, P. del Giudice, E. Tabet eds. ETS Editrice, Pisa (1991).

[97] Talagrand, M., *Concentration of Measure and Isoperimetric Inequalities in Product Spaces*, preprint Université Paris VI (1995).

[98] Talagrand, M., *A New Look at Independence*, preprint Université Paris VI (1995).

[99] Talagrand, M., *Résultats Rigoureux pour le Modèle de Hopfield*, preprint Université de Paris VI (1995).

[100] Tarkowski, W., M. Lewenstein, Storage of Sets of Correlated Data in Neural Network Memories, *J. Phys. A: Math. Gen.*, **26**, 2453–2469 (1993).

[101] Tishby, N., E. Levin, S. Solla, Consistent Inference of Probabilities in Layered Networks: Predictions and Generalisation, *IEEE Neural Net.*, **2**, 403–410 (1989).

[102] Vermet, F., Convergence de la Variance de l'Énergie Libre du Modèle de Hopfield, *C. R. Acad. Sci. Paris*, **315**, 1001–1004 (1992).

[103] Vermet, F., *Étude Asymptotique d'un Réseau Neuronal : le Modèle de Mémoire Associative de Hopfield*, PhD Thesis, Université de Rennes I, presented on 28 January (1994).

[104] Vermet, F., *Asymptotic Study of a Neural Network*, preprint Université de Rennes I (1994).

LOW COMPLEXITY AND GEOMETRY

GERARD RAUZY
Laboratoire de Mathématiques Discrètes
C.N.R.S. UPR 9016
163 Av. de Luminy, Case 930
13009 Marseille Cedex 9
France
rauzy@lmd.univ-mrs.fr

ABSTRACT. We give in the following a definition of the complexity function of an infinite sequence on a finite alphabet. This function measures in some sense the predictability of the sequence. It is defined as the counting function of blocks of length n occurring in the given sequence. These blocks can be arranged in graphs and we study in a peculiar case (sturmian sequences) the evolution of these graphs. It turns out that low complexity implies strong geometrical properties of the sequence viewed as the itinerary of a point in some dynamical system. This field of research is rich of yet unsolved problems.

1. What Could be a Reasonable Definition of Complexity ?

1.1. DEFINITIONS AND NOTATIONS

1.1.1. *Words.* Let $\mathbf{A}$ be a finite set that in the following we will call an alphabet and the elements of which letters.

Let E be a finite set which generally in the following will be the empty set or the set $\{0, \cdots, n-1\}$ where n is a strictly positive integer. Sometimes we will use also the set $\{1, \cdots, n\}$ where n is a strictly positive integer.

A word is by definition a map of E into $\mathbf{A}^*$.

The set of words on the alphabet $\mathbf{A}$ will be denoted by $\mathbf{A}^*$.

The empty word that is to say the word which corresponds to the case where E is the empty set will generally be denoted $1_{\mathbf{A}}$ or if there is no ambiguity on the alphabet by 1, or when the alphabet is a set of numbers by e.

The length of a word u of $\mathbf{A}^*$ denoted by $|u|$ is the cardinality of the set E.

So the only word of length 0 is the empty word and words of length 1 sometimes called singletons consist of one letter only: generally we will confound singletons and letters assimilating $\mathbf{A}$ to a subset of $\mathbf{A}^*$.

On $\mathbf{A}^*$ there is an important operation called concatenation or product which associates to two words u and v of $\mathbf{A}^*$ a word w denoted uv which is defined as follows:

- if u is the empty word, then

$$uv := v$$

- if v is the empty word, then

$$uv := u$$

- if $E = \{0, \cdots, m-1\}$, $F = \{0, \cdots, n-1\}$ and u, v are maps of respectively E, F into $\mathbf{A}$ then w is the map of $\{0, \cdots, m+n-1\}$ into $\mathbf{A}$ defined by

$$\begin{aligned} w(i) &:= u(i) \qquad \text{if} \quad i < m \\ w(i) &:= v(i-m) \text{ else} \end{aligned}$$

This enables us to give a structure of monoid on the set $\mathbf{A}^*$ where the empty word is the unitary element that is to say

$$1_{\mathbf{A}}u = u1_{\mathbf{A}} = u \quad \text{for every} \quad u \quad \text{in } \mathbf{A}^*$$

The product is associative:

$$(uv)w = u(vw)$$

so we can avoid parentheses and write this common value uvw.

This convention and the assimilation of $\mathbf{A}$ to the set of singletons permit for a word u of $E = \{0, \cdots, n-1\}$ to $\mathbf{A}$ the writing

$$u = u(0) \cdots u(n-1)$$

Remark. Those preliminaries could look like trivialities.

They are indeed, but in the combinatorics on words it is essential to distinguish a formal expression and its value which is generally also a formal expression.

1.1.2. *Factors, Prefixes, ...*. Let u be a word of $\mathbf{A}^*$. A factor v of u is a word of $\mathbf{A}^*$ such that there exists two words w, w' of $\mathbf{A}^*$ (possibly empty) such that

$$u = wvw'$$

A prefix of u is a factor v such that in the previous notation w is the empty word.

A suffix of u is dually a factor v where w' is empty.

We will generally denote by $Fact(u)$, $Pref(u)$ and $Suf(u)$ the sets of factors, prefixes and suffixes of u.

Evidently u belongs to $Pref(u)$. We will have oftenly to consider prefixes v of u that have a length strictly less than the length of u: we will call them *proper prefixes* of u and the corresponding set will be denoted by $Pref^*(u)$.

1.1.3. *Languages, Grammars, ...*. A set of words on a given alphabet will be called a language on this alphabet.

There are three basic types of operations on languages:

- boolean operations

These are the ordinary operations of set theory: union, intersection and complementation

- product, powers

Let L and M be two languages on the alphabet $\mathbf{A}$.

The product LM of these languages is the set of words w such that

$$w = uv \qquad \text{with } u \text{ in } L \text{ and } w \text{ in } M$$

The $n-th$ power of L denoted by L^n is defined by recurrence

$$- L^0 := 1_{\mathbf{A}}$$

$$- L^{n+1} := LL^n = L^nL$$

- star

The star of a language L is the union of all powers of L.

It is denoted by L^*.

Remark. This notation becomes coherent with the notation $\mathbf{A}^*$ for the set of all words on $\mathbf{A}$ once we have assimilated letters of $\mathbf{A}$ to singletons.

Remark. There are other operations on languages or words. One of them is the mirror image which consists to read words from right to left.

This recovers a notion of duality which is evident for instance when considering prefixes or suffixes and it will become apparent later. Nevertheless at that stage it seems to be unnecessary to introduce a special notation.

It is sometimes possible to describe a language by giving an algorithm which enumerates the words belonging to it. Such an algorithm, we will call it a grammar.

We do not want here to define precisely this notion, but just have it in your mind!

1.1.4. *Infinite Sequences, Shift, Associated Languages.* We will be concerned with one-sided sequences that is to say maps of the set $\mathbb{N}$ of non-negative integers into the alphabet $\mathbf{A}$, or two-sided sequences where $\mathbb{N}$ is replaced by the set $\mathbb{Z}$ of relative integers.

On each of this sets there exists a basic operation which is called the shift.

Let $\underline{x} = (x_n)$ be a sequence of letters of $\mathbf{A}$ indexed by $\mathbb{N}$ or $\mathbb{Z}$. In either cases the shift of x is the sequence $\underline{y} = (y_n)$ indexed by the same set such that

$$y_n := x_{n+1}$$

Whatever may be the alphabet (or the indexing set) we shall generally denote by T the map so defined and write

$$\underline{y} = T\underline{x}$$

Powers like T^n are standard notations.

We now define the language $Fact(\underline{x})$ associated to an infinite sequence $\underline{x}$.

- the empty word belongs to $Fact(\underline{x})$

- the word $u = u_0 \cdots u_{m-1}$ belongs to $Fact(\underline{x})$ if there exists an integer n (element of $\mathbb{N}$ or $\mathbb{Z}$ according to the set indexing $\underline{x}$) such that

$$u_i = x_{n+i} \qquad \text{for every} \quad i = 0, \cdots, m-1$$

Observe that

- if u belongs to $Fact(\underline{x})$ then every factor of u belongs also to $Fact(\underline{x})$ (such a language is called factorial)

- if u belongs to $Fact(\underline{x})$ then there exists a letter a such that ua belongs to $Fact(\underline{x})$. If $\underline{x}$ is a two-sided sequence there also exists a letter b such that bu belongs to $Fact(\underline{x})$. One-sided sequences with the same property are called recurrent. It can easily be verified that if a word u of length m belongs to $Fact(\underline{x})$ where $\underline{x}$ is recurrent then there exists infinitely many n such that

$$u_i = x_{n+i} \qquad \text{for every} \quad i = 0, \cdots, m-1$$

1.2. SEQUENCES AND WAYS TO LOOK AT

1.2.1. *Random Sequences.* Cast a dice or a coin and look at the sequence obtained...

Conversely does the sequences $HTTTHHTTHHHTHT\ldots$ seems to be a "normal" series playing heads and tails?

Here evidently, we can only observe statistical laws: if the play is fair, the frequencies of occurrences of a given word, say for instance HT, must be around the mathematical expectation - in that case $1/4$ - and this has to be true for any word.

Probability theory asserts that such sequences exist and that in some measure-theoretical sense almost all sequences are like that.

But there exist also algorithms to construct "deterministic" sequences which verify the law about frequencies. The best known example in Champernowne's sequence: write the sequence of non-negative integers 0, 1, 2, ... in the binary system of numeration. We get the sequence

$$011011100101110111\ldots$$

and it can be proved that such a sequence verifies the statistical laws about frequencies of occurring words.

Is that sequence a complex one?

The algorithm which generates it, seems to be very simple.

But suppose you have a computer with a finite memory. You have written the first n terms of the sequence, now going forwards or backwards, you can read this beginning and decide to do so according with some rule which at a time ends and implies that the $n+1$ - th term is 0 or 1. Do you think that it is possible to find such a rule for the Champernowne's sequence?

Can you construct a sequence which is not a periodic one or ultimately periodic with such a prescription?

If $\underline{\mathrm{x}}$ is the sequence of digits of the square root of 2, what can be said about its statistical behaviour.

1.2.2. *A Typical Example.* Let $\underline{x}$ be a given one-sided sequence.

Define a function g of $\mathbb{N}$ into $\mathbb{N}$ by

$$- \, g(0) := 0$$
$$- \, g(n+1) := \inf\{k / x_k \neq x_{g(n)}\}$$

The function g is well defined unless there exists n such that

$$x_k = x_n \qquad \text{for every} \quad k \geq n$$

Now the sequence of "runs" in $\underline{\mathrm{x}}$ is the sequence $\underline{r}$ of positive integers such that

$$r(n) := g(n+1) - g(n)$$

For instance let $\underline{x}$ be the sequence which begins by

$$\underline{x} = 22112122122112\ldots$$

then

$$\underline{r} = 22112122\ldots$$

The sequence r has the same beginning as the sequence x and it is easy to see that there exists only one sequence $\underline{x}$ on the alphabet $\{1,2\}$ with that property.

It can be produced by a very simple machine (called a transducer) in the following way:

- the machine has two states called s and t, the initial state being s
- inputs and outputs are words on the alphabet $\{1,2\}$
- we read the input from left to right letter by letter
- if the machine is in the state s and we read 1 we write 2, if we read 2 we write 22 and in both cases the machine goes to the state t
- similarly, if the machine is in the state t and we read 1 we write 1, if we read 2 we write 11 and in both cases the machine goes to the state s.

Example. Start with the word 122

- we read 1, the machine is in the initial state s we write 2
- now we read the second letter of the word which is 2 and we know that the machine is in the state t, so we write 11
- the next and last letter is 2 again but the machine is in the state s, so we write 22.

So the output is the word 21122.

This process can easily be extended to infinite one-sided sequences, so it defines a mapping from $\{1,2\}^{\mathbb{N}}$ into itself. It is quite trivial that there exists only one fixed point (starting from the letter 2 and the state s) and this is precisely the sequence $\underline{x}$ we looked at.

Make a program which writes the first million of digits of x.

Now what can be said about the sequence?

As far as I know, quite nothing.

It is not a periodic one (or ultimately periodic): try to prove it.

The upper density of $1'_s$ and $2'_s$ is bounded by 0.5006. A reasonable conjecture would be that there exists a density equal to $1/2\ldots$

Suppose we do the same thing with two letters 1, 3 starting by 3. We get the sequence

$$\underline{z} = 333111333131313131333\ldots$$

Surprisingly that sequence is simpler than the preceding one and statistics can be made.

1.2.3. *Self-Referent Sequences.* The previous examples are typically definitions of a sequence referring to itself: we write the first n terms, and to write the $n+1$-th we have to look back to the first m terms where m is less than n.

Self-reference is at the beginning of logical problems: we have no clear answer to the question "why $\underline{z}$ is simpler than $\underline{x}$?".

There exists a definition of the complexity of a finite sequence due to Kolmogorov and revised by Chaitin.

Roughly speaking, suppose the given sequence is on the alphabet $\{0, 1\}$.

We try to do a program to write its successive terms: this program can also be written as a sequence on the same alphabet. The shorter this sequence is, the smaller is the complexity of the given sequence (precise definitions can be given).

In that sense, a random sequence is defined as a sequence for which the shortest program to write it has about the same length as the sequence: this is quite satisfactory.

But this definition is not "effective": given a sequence there doesn't exist an algorithm which writes the shortest program which produces that sequence. The reason is precisely to be find in self-referent sequences.

There are many constructions of sequences: self-reference is the simplest way to construct a sequence which is not an ultimately periodic one.

In what follows, we shall restrict ourself to a special kind of self-reference called substitution. We also will look at a rough measure of the complexity of such a sequence. It will appear that there exists some consistency within those notions and that geometrical interpretations are useful to understand this observation.

1.3.1. *Fibonacci Sequence.* Let $A = \{0, 1\}$ and f the function of $\mathbf{A}$ into $\mathbf{A}^*$

$$f(0) := 01$$
$$f(1) := 0$$

We extend this definition to words and sequences on $\mathbf{A}$

$$f(ax) := f(a)f(x)$$

Starting from the word 0 we get successively

$$01$$
$$010$$
$$01001$$
$$\dots$$

It is easy to see that if $u(n)$ is the word obtained at the $n - th$-state $u(n)$ is a prefix of $u(n + 1)$.

We can construct a self-referent sequence called Fib

$$Fib := 0100101001001\dots$$

which begins by $u(n)$ for every n.

Evidently, it verifies

$$Fib = f(Fib)$$

and it is the only one sequence with this property.

1.3.2. *A Geometrical Interpretation.* Define $X := [0, 1[, X_0 := [0, \alpha[, X_1 := [\alpha, 1[$ where $\alpha := (\sqrt{5} - 1)/2$

Now define a map T of X into itself by

$$T(x) := x + 1 - \alpha \quad \text{if} \quad x \in X_0$$
$$:= x - \alpha \quad \text{if} \quad x \in X_1$$

Let $\nu(x)$ the element of $\{0,1\}$ such that x belongs to $X_{\nu(x)}$. The sequence $(\nu(T^n x))_{n\in\mathbb{N}}$ is called the itinerary of x.

Now we have the following properties.

Proposition 1.3.2.

1 - For every x in X, the orbit of x namely the sequence $(T^n(x))_{n\in\mathbb{N}}$ is dense in X.

2 - If x is a point in X, $\underline{\text{it}}(x)$ its itinerary the language $Fact(\underline{\text{it}}(x))$ associated as in 14, doesn't depend of x. We will denote it by L

3 - The number $p(n)$ of elements of L of length n is

$$p(n) = n + 1$$

4 - The sequence Fib is the itinerary of the point α^2.

Remark. 1, 2, 3 are true for every irrational number $\alpha \in [0,1[$

Proofs of 1 and 2 are left to the reader.

To prove 3, first observe that by virtue of 1 and 2, a word

$$u = u(0) \cdots u(n-1)$$

belongs to L if and only if there exists $x \in X$ such that

$$\nu(T^k x) = u(k) \qquad \text{for } k = 0, \cdots, n-1$$

Now we can conclude using the lemma.

Lemma. Let E_n be the set of fractional parts $\{k\alpha\}$ for $k = 1, \cdots, n$.

And suppose $E_n = \{\beta_1, \cdots, \beta_n\}$ where

$$0 < \beta_1 < \ \cdots \ < \beta_n < 1$$

Then denoting 0 by β_0, 1 by β_{n+1} we have

$$(i) \qquad \forall\, k = 0, \cdots, n \quad \forall\, x, y \in [\beta_k, \beta_{k+1}[$$

$$\nu(x)\,\nu(Tx)\;\cdots\;\nu(T^{n-1}x)=\nu(y)\,\nu(Ty)\;\cdots\nu(T^{n-1}y)$$

(ii) If $h\neq k,\ x\in[\beta_h,\beta_{h+1}[,\ y\in[\beta_k,\beta_{k+1}[$

$$\nu(x)\,\nu(Tx)\;\cdots\;\nu(T^{n-1}x)\neq\nu(y)\,\nu(Ty)\;\cdots\nu(T^{n-1}y)$$

The property is true for $n=1$.

Assume that is true for a given n, we want to prove it for $n+1$. Observe first that T is a bijective transformation and that the inverse transformation T^{-1} is defined in the same way as T by exchanging the two intervals $y_0=[0,1-\alpha[,\ y_1=[1-\alpha,1[$.

Now let $h=0,\cdots,n$ two cases can occur

(i) if $1-\alpha\notin[\beta_h,\ \beta_{h+1}[$ then $T^{-1}[\beta_h,\ \beta_{h+1}[$ is again an interval shifted by a translation of α mod 1 from the interval $[\beta_h,\ \beta_{h+1}[$, so its extremities belong to the set $E_{n+1}=\{\gamma_1,\cdots,\gamma_{n+1}\}$

(ii) if $1-\alpha\in[\beta_h,\ \beta_{h+1}[$ it must be an interior point which divides $[\beta_h,\ \beta_{h+1}[$ in two intervals.

$$\begin{array}{llll}\text{By } T^{-1} & [\beta_h,1-\alpha[& \text{goes to} & [\gamma_{n+1},[\gamma_{n+2}[\\ & [1-\alpha,\beta_{h+1}] & \text{goes to} & [\gamma_0,[\gamma_1[\end{array}$$

Suppose x,y are two points of $T^{-1}[\beta_h,\beta_{h+1}[$.

In case (i)

$T^{-1}[\beta_h,\beta_{h+1}[$ is included in X_0 or X_1, so $\nu(x)=\nu(y)$ and by the recurrence property

$$\nu(Tx)\cdots\nu(T^nx)=\nu(Ty)\cdots\nu(T^ny)$$

In case (ii)

- either there are both in the same interval $[\gamma_0,\gamma_1[$ or $[\gamma_{n+1},\gamma_{n+2}[$ and again

$$\nu(x)\nu(Tx)\cdots\nu(T^nx)=\nu(y)\nu(Ty)\cdots\nu(T^ny)$$

- or there are in two different ones and

$$\nu(x)\nu(Tx)\cdots\nu(T^n x) = \nu(y)\nu(Ty)\cdots\nu(T^n y)$$

The proof of the lemma follow easily.

We defer the proof of 4 to the following chapter.

1.4. COMPLEXITY

1.4.1. *Fibonacci Again.* Fibonacci sequence is perhaps the simplest example of a self-referent sequence. We could think that the notion of complexity may take this fact in account.

Proposition 1.3.2 seems to indicate that the function $p(n)$ plays a role to describe the complexity.

Furthermore, proposition 1.3.2 proves that such a function can "easily" be computed when some geometrical properties (i.e 4) are known.

For instance we looked here at transformations which exchange two intervals. A bit more complicated argument can be extended to exchanges of n intervals.

Another remark is that starting with a combinatorial definition of Fibonacci sequence (with a substitution), we first give a geometrical interpretation of it (the step 4 in proposition 1.3.2) to return to a combinatorial property about the number of factors of length n.

1.4.2. *The Complexity Function.*

Definition. Let $\underline{x}$ be a sequence, $L = Fact(\underline{x})$ the language associated to $\underline{x}$. The complexity function of $\underline{x}$ denoted by $p(\underline{x}, n)$ is the number of factors of length n in L.

Remark. Instead of speaking of the complexity function of a sequence, we could also speak of the complexity of a language ...

1.4.3. *Complexity and Prediction.* Let $\underline{x}$ be any sequence on the alphabet A and $p(n) = p(\underline{x}, n)$ its complexity.

Suppose you know that a word of length n say $u = u(0)\cdots u(n-1)$ occurs at some place in the sequence i.e. there exists k such that

$$\underline{x}(k)\cdots\underline{x}(k+n-1) = u$$

Problem. Make a prediction about the letter $\underline{x}(k+n)$.

Define the function ∂^+ on words of $Fact(\underline{x})$ by

$$\partial^+(u) = {}^{\#}\{a \in A \mid ua \in Fact(\underline{x})\}$$

If $\partial^+(u) = 1$ there exists an only one possible continuation say ua and we can make an exact prediction

$$\underline{x}(k+n) = a$$

So there are two extreme cases

(i) for every word u of length n $\qquad \partial^+(u) = 1$

(perfect prediction)

(ii) for every word u of length n $\qquad \partial^+(u) = {}^{\#}A$

(no possible prediction)

As a word w of length $n+1$ in $Fact(\underline{x})$ can in an unique namer be written $w = ua$ we have

$$\sum_{u \in L_n} \partial^+(u) = p(n+1)$$

(here L_n is the set of words of length n in $Fact(\underline{x})$).

Remark that if $u \in L_n$ there exists at least a letter $\underline{a}$ such that $ua \in L_{n+1}$. In other terms $\partial^+(u) - 1$ is a positive integer and we have a fundamental equation

$$\sum_{u \in L_n} (\partial^+(u) - 1) = p(n+1) - p(n)$$

In case (i) $p(n+1) - p(n) = 0$ $\qquad$ i.e. $p(n+1) = p(n)$

In case (ii) $p(n+1) - p(n) = p(n) \times ({}^{\#}A - 1)$ $\qquad p(n+1) = {}^{\#}Ap(n)$

Evidently if case (ii) is true for all n, we get

$$p(n) = ({}^{\#}A)^n \text{ so that } \lim_{n\to\infty} \frac{1}{n}\, p(n) = {}^{\#}A$$

It is more interesting to look carefully at the case (i).

Suppose it occurs for some n.

Then if a word v belongs to L_{n+1}, there exists a letter $\underline{b}$ of A and a word u of L_n such that

$$v = bu$$

So
$$\partial^+(v) \le \partial^+(u)$$

and thus
$$\partial^+(v) = 1$$

We can conclude that

$$\sum_{v \in L_{n+1}} (\partial^+(v) - 1) = 0 = p(n+2) - p(n+1)$$

and by recurrence

$$p(m) = p(n) \qquad \text{for every} \quad m \ge n$$

Exercise. Show that this can happen if and only if $\underline{x}$ is ultimately periodic.

We can conclude at a first look that the growth of the sequence $p(n+1) - p(n)$ measures in a certain sense the ability to "predict" the sequence $\underline{x}$.

Remark. We can also define in the same way a function ∂^- on $Fact(\underline{x})$.

In the case where $\underline{x}$ is recurrent (see definition in 1.4), we have the same formula:

$$\sum_{u \in L_n} (\partial^+(u) - 1) = p(n+1) - p(n)$$

Exercise. Show that the quantity $\frac{1}{n}$ log $p(n)$ has a limit when n goes to infinity.

This limit is called the <u>topological entropy</u> of the dynamical system associated to $\underline{x}$.

If this limit is 0, we will say that $\underline{x}$ is a <u>deterministic</u> sequence. But we will not make an extensive use of that notion.

1.4.4. *Low Complexity and Geometry.* We said in 4.3 that if for a sequence $\underline{x}$, the function $p(n)$ is not strictly increasing then $\underline{x}$ is ultimately periodic.

So if we want to get a sequence $\underline{x}$ which is not an ultimately periodic one, we must suppose that $p(n+1) - p(n) \geq 1$ for every n, and the lowest case is :

$$p(n) = n + 1$$

As we know, Fibonacci sequence verifies such a property and this can be proved by some geometrical argument. We said that this argument can be extended to many other sequences obtained from exchanges of two intervals.

But the converse is true:
"if a sequence $\underline{x}$ is such that $p(n) = n + 1$, then it has a geometrical interpretation as an itinerary in some 2 interval-exchange".

This is a very old observation (not exactly expressed in the same terms) and one question arises:
"if a sequence $\underline{x}$ verifies a simple formula for $p(n)$, has it a geometrical interpretation ?"

From a long time ago, we looked at sequences verifying

$$p(n) = 2n + 1 \qquad \text{i.e.} \qquad p(n+1) - p(n) = 2$$

At the beginning of that story, it seemed to be the simplest generalization because:

(i) we had easy examples with a geometrical interpretation

(ii) generalizations to $p(n+1) - p(n) = s$ seemed to be easy.

Nowadays even that question is not solved in its full generality. Combinatorial conditions must be added to obtain necessary and sufficient characterizations.

But many progresses occurred and we will try to make a survey of them.

The conclusion we could draw is that the function $p(n)$ is not sufficient to really describe the complexity of a sequence.

To get a geometrical representation needs some additional combinatorial properties.

Geometry is far from inducing clear combinatorial properties.

Furthermore "geometry" has to be taken in an extended sense including geometry on ultrametric spaces for instance.

Nevertheless, there is something behind that, we can only understand at that time, step by step.

Real understanding of the whole will come from such an addition of results, conditioning the ways we ask ourself the questions.

2. Billiards, Substitutions and Automata

2.1 INDUCED MAPS

2.1.1. *Definition.* Let (X, S) be a couple where X is a given space and S a map from X into itself.

Let Y be a subspace of X and x a point of Y.

The first return time $\tau(x)$ is defined as follows

$$\begin{aligned}\tau(x) &= \inf\{n > 0 \mid S^n(x) \in Y\} \\ &= \infty \text{ if } \{n > 0 \mid S^n(x) \in Y\} \text{ is empty.}\end{aligned}$$

The first return map, or the induced map of S over Y is the map T of $Y/\{x \in Y \mid \tau(x) = \infty\}$ into Y such that

$$T(x) = S^{\tau(x)}(x)$$

Poincaré's recurrence lemma asserts that if there exists on X a measure μ which is S-invariant, i.e.

for every measurable set $E \subset X$

$$\mu(S^{-1}E) = \mu(E)$$

and if $Y \subset X$ is any measurable set then T is defined almost everywhere on Y that is to say

$$\mu(\{x \in Y \mid \tau(x) = \infty\}) = 0$$

2.1.2. *Interval Exchanges.* We generalize the notion introduced in Chapter 1 - par. 3.2.

X is any interval $[a, b[$ of the line ($a, b \in \mathbb{R}$), and X_0, X_1 a partition of X in two intervals $[a, c[$, $[c, b[$.

We assume that the ratio of length $\alpha = |X_0|/(|X_0| + |X_1|)$ is an irrational number, and denote by S the transformation which exchanges X_0 and X_1 that is to say

$$\begin{aligned} Sx &:= x + b - c \quad \text{if} \quad x \in [a, c[\\ &:= x + a - c \quad \text{if} \quad x \in [c, b[\end{aligned}$$

Now take $Y = X_0$.

Then, it is easy to prove that the induced map of S over Y is everywhere defined and is again a $2-$interval exchange

The figure represents the case where $|X_0| > |X_1|$,

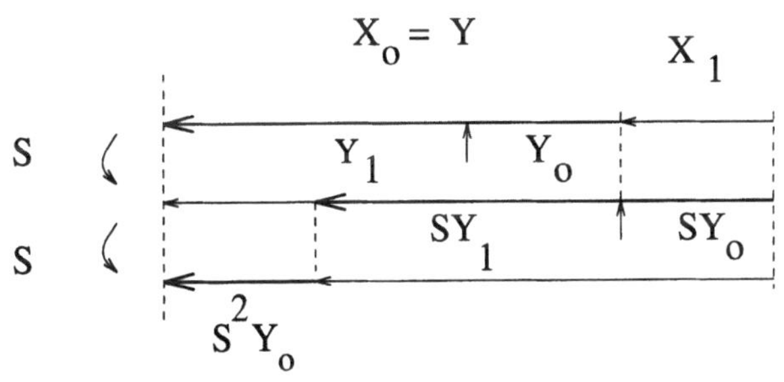

Figure 1

(For technical reasons we reverse the sense for the partition of Y in the two intervals Y_0, Y_1).

Observe that

$$\begin{aligned} |X_0| &= |Y_0| + |Y_1| \\ |X_1| &= |Y_0| \end{aligned}$$

In the general case

$$\begin{aligned} |X_0| &= |Y_0| + |Y_1| \\ |X_1| &= |Y_0| + k(|Y_0| + |Y_1|) \end{aligned}$$

where $k = [1/\alpha] - 1$.

Now if we look at the ratio $\beta = \frac{|Y_0|}{|Y_0|+|Y_1|}$ we get the formula

$$\alpha = \frac{1}{a+\beta} \quad \text{where} \quad a = [1/\alpha]$$

This is a well-known formula which is called continued – fraction expansion.

For example, if $\alpha = (\sqrt{5}-1)/2$ then $a = 1$, $\beta = \alpha$ so the process iterates and it is generally written

$$\alpha = 1/1 + 1/1 + \dots$$

To summarize, the map $S \longmapsto T$ which associates to S the induced map of S over Y gives a geometrical interpretation of that process.

2.1.3. *Itineraries.* For $x \in X$ define $\mu(x) = i$ if $x \in X_i$.

Similarly for $x \in Y$, define $\nu(x) = i$ if $x \in Y_i$.

What are the relations for a given point x of Y between the two sequences (itineraries):

$$\underline{u} = \mu(x)\mu(Sx)\cdots\mu(S^n x)\cdots$$

$$\underline{v} = \nu(x)\nu(Tx)\cdots\nu(T^n x)\cdots$$

Suppose for instance that we are in the case of figure 1.

We know that by S, Y_0 first goes to X_1 and then comes back into Y so that on Y_0 we have $T = S^2$, while Y_1 comes back immediately into Y so that on Y_1 we have $T = S$.

Suppose now that $\underline{v}$ begins by $010\dots$

We have

$v_0 = 0$	so	$x \in Y_0 \subset X_0$	$\mu(x) = 0$
		$Sx \in SY_0 \subset X_1$	$\mu(Sx) = 1$
		and $S^2x = Tx$	

now

$v_1 = 1$	so	$S^2x \in Y_1 \subset X_0$	$\mu(S^2x) = 0$
		and $S^3x = T^2x$	

$v_2 = 0$ so $S^3x \in Y_0 \subset X_0$ $\mu(S^3x) = 0$

$S^4x \in SY_0 \subset X_1$ $\mu(S^4x) = 1$

and $S^5x = T^3x$

....

In other terms, the sequence u is obtained by applying to v the substitution f we considered in Chapter 1. :

$$\underline{u} = f(\underline{v}) \qquad \text{where} \qquad \begin{cases} f(0) = 01 \\ f(1) = 0 \end{cases}$$

Exercise. Prove now part 4 of the proposition 1.3.2 in Chapter 1.

2.2. BILLIARDS

2.2.1. *Billiard in the Square.*

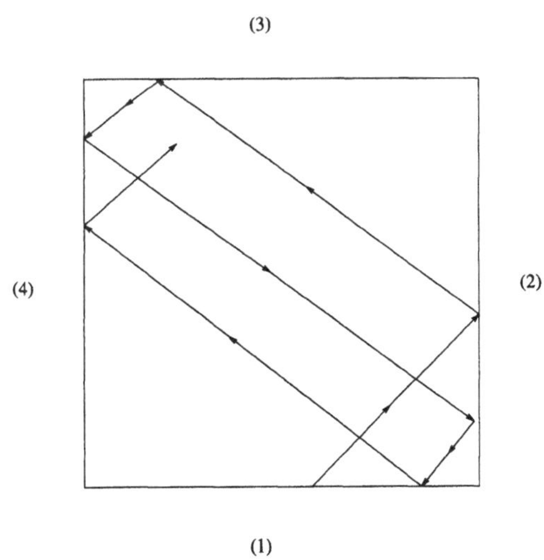

Figure 2

A ball start from the first side (1) in a given direction.

Except particular cases where it goes in a corner, it attains another side where it reflects according to the ordinary laws.

We are only interested by the sequence of successive sides, its trajectory meets.

In the case of figure 2 this sequence begins by:

$$(1)\ (2)\ (3)\ (4)\ (2)\ (1)\ (4)\ \ldots$$

The general problem is to give a combinatorial characterization of such a sequence.

2.2.2. *Unfolding a Billiard.* Instead of reflecting the ball, we can also consider that it goes along a straight line and reflect the square itself. For instance figure 3 shows two consecutive reflections

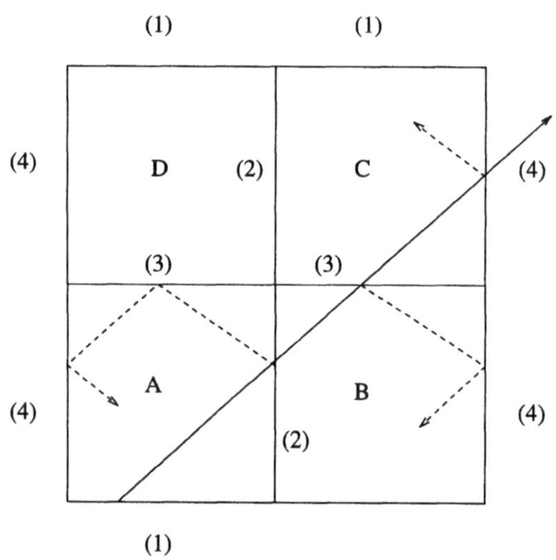

Figure 3

Reflected on the side (2) the original square A becomes the square B, and reflected B on its side (3) gives the square $C \ldots$

We have only four types of squares

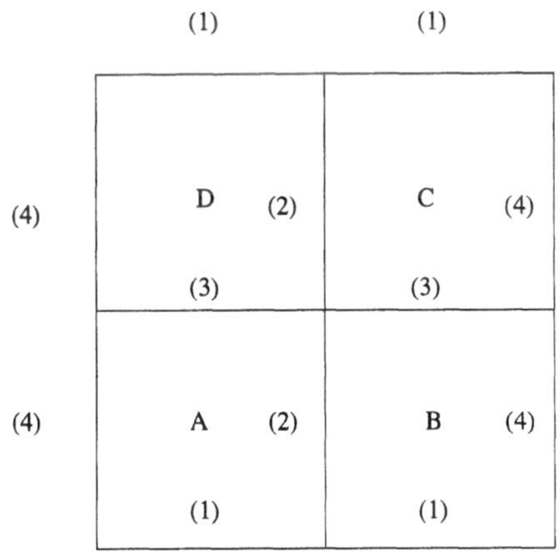

Figure 4

Now look at a sequence $\underline{x}$ obtained in such a way, and replace (1) and (3) by a, (2) and (4) by b.

In our example, we get a sequence $\underline{y}$

$$\underline{y} = a\ b\ a\ b\ b\ a\ b\ \dots$$

Question. Knowing the sequence $\underline{y}$ can we recover the sequence $\underline{x}$?

The answer is yes.

First we know that we are starting from side (1) of a square of type A.

Now to go out of that square we have 2 possibilities:

- either, we enter in B, the second lister of $\underline{y}$ is then b and the second letter of $\underline{x}$ is (2)

- or, we enter in C, the second letter of $\underline{y}$ is then a and the second letter of $\underline{x}$ is (3).

We have to examine the four cases entering in A, B, C, D and going to one another type which is evidently ruled by the sequence $\underline{y}$. Those rules can be represented by a transducer in the sense of Chapter 1.

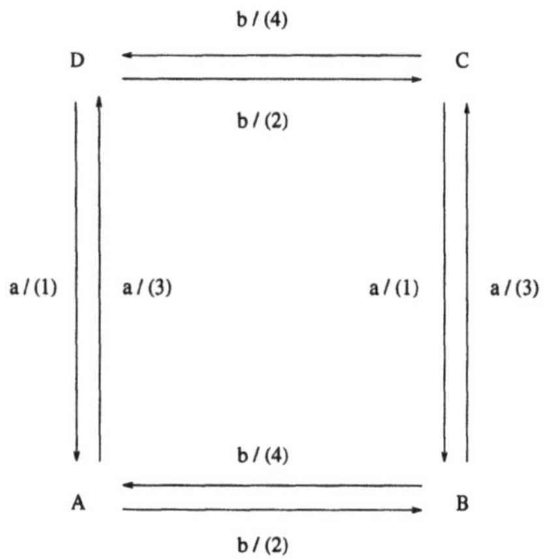

Figure 5

the original state being D.

As a matter of fact this transducer is simpler than the one we encountered in Chapter 1 - § 2.2 :

(i) when we read a letter, we write a letter only

(ii) there are evident symmetries.

Those facts permit to compute the complexity function of $\underline{x}$ knowing the complexity function of $\underline{y}$.

Try do do so!

2.2.3. *Billiards and Interval Exchanges.* We are know only interested by sequences such as $\underline{y}$ on the alphabet $\{a, b\}$ obtained by intersecting a square grid by a straight line Δ.

If Δ, has a rational slope such a sequence is periodic and it is very interesting to understand the structure of words w (called sturmian words) for which

$$\underline{y} = w\ w\ w\ \ldots$$

But in the following, our main interest is to look at the case where the slope is an irrational number λ.

Suppose the line starts from a point x of the diagonal of slope - 1 of a square A, there are two cases

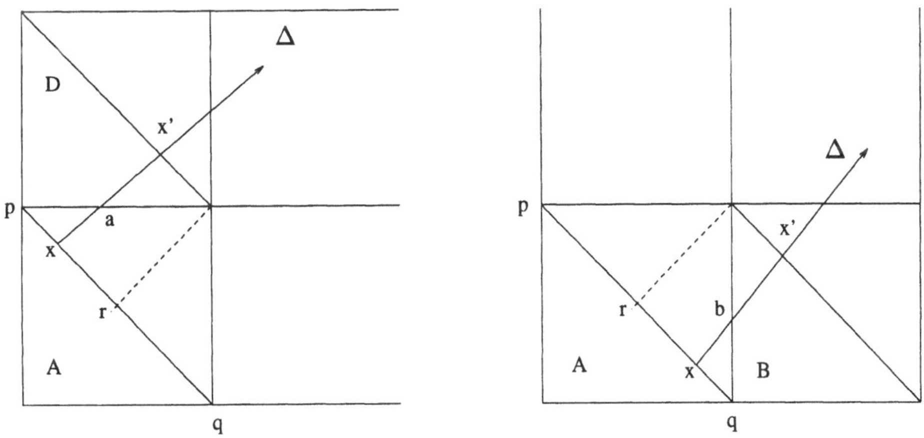

Figure 6

Then it crosses in case (1) a side of A named a, enters in D, and crosses the diagonal of D at a point x', in case (2) the side of A is named b, the line enters in B and crosses the diagonal of B at x'.

In any case the knowledge of x' determines the following terms of $\underline{y}$. More precisely, set

$$\nu(x) = a \quad \text{if} \quad x \quad \text{belongs to the segment} \quad pr$$

$$\nu(x) = b \quad \text{if} \quad x \quad \text{belongs to the segment} \quad rq$$

(we neglect the extremities).

Then

$$\underline{y}(x) = \nu(x)\ \underline{y}(x')$$

(This looks like an induced map and it is but for a mathematical object called a flow that we have not defined).

Denote by T the transformation of pq into itself (identifying all diagonals) such that $x' = Tx$.

We see on Figure 6 that

$$\begin{aligned} x' &= x + \overrightarrow{rq} \qquad \text{if} \qquad x \in pr \\ &= x + \overrightarrow{qr} \qquad \text{if} \qquad x \in rq \end{aligned}$$

So T defined on pq is exactly the exchange of the two intervals pr and rq and the ratio of lengths is

$$|pr|/|pq| = \lambda/(1+\lambda)$$

We are exactly in the situation of Chapter 1 - § 3.2 : $\underline{y}$ is the itinerary of x, i.e.

$$\underline{y} = \nu(x)\ \nu(Tx)\ldots : \nu(T^n x)\ldots$$

By proposition 1.3.2, the complexity of $\underline{y}$ is

$$\mathbf{p(n) = n + 1}$$

2.2.4. *Fibonacci Sequence Again.* Evidently starting from a 2-interval exchange on the interval X we can always consider that X is the diagonal of a square and apply the same construction.

So Fibonacci sequence can be obtained from a square billiard where λ is the golden ratio: $\lambda = (\sqrt{5}+1)/2$.

2.3. GRAPHS OF WORDS

2.3.1. *Introduction.* We have in mind to prove that in some sense the converse is true, that is to say if a sequence has a complexity function

$$p(n) = n + 1$$

then it can be obtained by a billiard sequence.

But we have to be cautious on the hypothesises.

For instance 2-sided sequences like

$$\ldots\ 0\ \ldots\ 0|1|0\ \ldots\ 0\ \ldots$$

or $$\ldots\ 01\ \ldots\ 01|00|01\ \ldots\ 01\ \ldots$$

verify that property and cannot be obtained as billiard sequences.

We do not give now a precise statement but it is necessary to introduce some tools which play an essential role.

2.3.2. *Definitions.* We start from a given language L on $\{a, b\}$.

We denote by L_n the set of words of L of length n.

We suppose that

(i) L is factorial

(ii) L is transitive, i.e. if u and v are words in L then there exists a word w such that, u is a prefix of w and v a suffix of w

(iii) the complexity function of L is not bounded.

Remark. (ii) and (iii) imply that for any word A in L, there exist letters c and d such that cA and Ad belong to L.

The graph $\Gamma_n(L)$ is an oriented graph defined as follows

- vertices are the elements of L_n

- if $A, B \in L_n$ and if there exists $C \in L_{n+1}$ such that A is a prefix of C and B a suffix of C then (A, B) is an edge of $\Gamma_n(L)$.

Example. Suppose L is the language associated to the Fibonacci sequence.

There are 3 words of length 2, namely 00, 01, 10, and 2 words of length 1, namely 0, 1.

So Γ_1 is the graph:

(we draw an arrow between A and B whenever (A, B) is an edge of the graph.

If (A, B) is an edge of $\Gamma_n(L)$, C is evidently uniquely determined: we have $C = Ab$ where b is the last letter of B.

We label $\Gamma_n(L)$ writing this letter upper the arrow

$$A \xrightarrow{b} B$$

So, Γ_1 for the Fibonacci sequence becomes

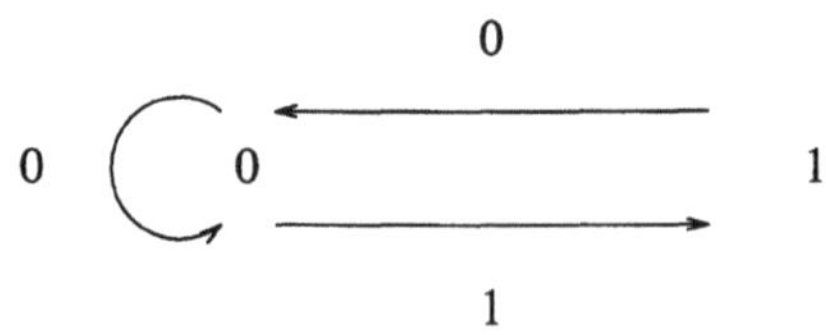

If in $\Gamma_n(L)$ we have a chain,

$$A_0 \xrightarrow{a_0} A_1 \xrightarrow{a_1} \dots \xrightarrow{a_{s-1}} A_{s-1}$$

its label is by definition the word $a_0\ a_1\ \dots\ a_{s-1}$

2.4. PRIMITIVE CHAINS

2.4.1. *Primitive Chains.* We say that $A_0 \xrightarrow{a_0} A_1 \xrightarrow{a_1} \dots \xrightarrow{a_{s-1}} A_s$ is a primitive chain if

(i) $\partial^+(A_0) = \partial^+(A_s) = 2$

(ii) $\partial^+(A_i) = 1 \quad \text{for} \quad i = 1, \dots, s-1$

n-names are the labels of primitive chains in Γ_n.

(we suppose here that we are in the case of a language L defined on $\{a, b\}$ as in § 4.2. Evidently, definitions are to be changed if the cardinality of the alphabet is greater than 2).

Remark. n-names belong to the language L and every word of L is a factor of an n-name for some n.

2.4.2. *From* n *to* $n+1$. Suppose that a word A in L_n verifies:

(i) $\partial^+(A) = 2$

(ii) $\partial^-(A) = 1$

Then there are 2 primitive chains starting from A

$$A \xrightarrow{a} A_1 \xrightarrow{a_1} \dots \xrightarrow{a_{s-1}} A_s$$

$$A \xrightarrow{b} B_1 \xrightarrow{b_1} \dots \xrightarrow{b_{t-1}} B_t$$

By hypothesis (ii) there also exists only one letter c such that cA is a word of L_{n+1}. Without restricting generality we can suppose that $c = a$.

Moreover we know that Aa, Ab, aA are in L, but bA isn't.

By the hypothesis of transitivity, aAa and aAb are also in L: starting from $u = aA$, we know that there exists a word w in L with prefix u and suffix Aa.

If there exists such a word of length strictly greater than $1 + |A|$, then we know that there exists a letter c such that cAa is a suffix of w. By condition (ii) $c = a$.

If the only word w is $w = aA$, then we have

$$aA = Aa$$

i.e. $A = \underbrace{a \dots a}_{n}$. But starting from aA, we could never get the word Ab.

Proof that $aAb \in L$ is easier.

Now we can say that

$$A' \xrightarrow{a} A'_1 \xrightarrow{a_1} \dots \xrightarrow{a_{s-1}} A'_s$$

$$A' \xrightarrow{b} B'_1 \xrightarrow{b_1} \dots \xrightarrow{b_{t-1}} B'_t$$

where $A' = aA$,

$$A'i = A_{i-1}A_i \text{ for } \quad i = 1, \dots, s$$
$$B'_i = B_{i-1}B_i \text{ for } \quad i = 1, \dots, t$$

are beginnings of primitive chains.

We do not know if they are primitive chains themselves because "a priori" we do not know what are $\partial^+(A'_s)$ or $\partial^+(B'_t)$.

The situation becomes much more complex if we suppose that

(i) $\qquad \partial^+(A) = 2$

(ii) $$\partial^-(A) = 2$$

Nevertheless, it can be fully analyzed.

Evidently, the situation depends of the number of A which $\partial^+(A) = 2$ and of an underlying graph (some simplified version of Γ_n labelled by n-names instead by letters). It seems that in that kind of things we have also to introduce words A such that $\partial^-(A) = 2$.

The whole story is not fully understood at that time.

So, in what follows, I will do a rough simplification, namely that there exists only one word A in L_n such that $\partial^+(A) = 2$.

2.5. $p(n+1) - p(n) = 1$ FOR EVERY n

There exists one word only, we call D_n such that $\partial^+(D_n) = 2$ so they are only 2 primitive chains

$$\begin{cases} D_n \xrightarrow{a} E_{n,1} \xrightarrow{e_1} \dots \xrightarrow{e_{s-1}} E_s = D_n \\ D_n \xrightarrow{b} F_{n,1} \xrightarrow{f_1} \dots \xrightarrow{f_{t-1}} F_t = D_n \end{cases}$$

(e_{s-1} is the last letter of E_s, and f_{t-1} the last one of F_t, so that $e_{s-1} = f_{t-1}$).

Now we look carefully at what happens going from n to $n+1$ in the same spirit as in § 5.2.

Evidently we have also an unique word G_n such that $\partial^-(G_n) = 2$.

For every n, we have two n-names

$$\begin{cases} U_n = & a\, e_1 & \dots & e_{s-1} \\ V_n = & b\, f_1 & \dots & f_{t-1} \end{cases}$$

and we start from $U_0 = a$, $V_0 = b$.

The result is:

Proposition 2.5.1.

(i) if $G_n \neq D_n$ then

$$\begin{cases} U_{n+1} = & U_n \\ V_{n+1} = & V_n \end{cases}$$

(ii) if $G_n = D_n$ then

either $\quad (0)\begin{cases} U_{n+1} = & U_n\, V_n \\ V_{n+1} = & V_n \end{cases} \quad$ or $\quad (1)\begin{cases} U_{n+1} = & U_n \\ V_{n+1} = & V_n\, U_n \end{cases}$

As we noticed every word in L is a factor of U_n or V_n for n large enough.

If at a time, one only of the rules (0) or (1) could be applied then it is easy to see that the complexity function of L would be bounded, contradicting (iii) in § 4.2.

We can now state a fundamental proposition.

Proposition 2.5.2. If L is a language verifying properties (i), (ii), (iii) of § 4.2, and if its complexity function p is such that

$$p(n) = n + 1$$

then there exists an infinite sequence $\underline{\epsilon} = \epsilon_0\ \epsilon_1, \ldots \epsilon_n\ \ldots$ with $\epsilon_k \in \{0,1\}$ such that

(i) $\quad \epsilon_k = 0$ and $\epsilon_k = 1$ occur for infinitely many values of k.

(ii) $\quad U_0 = a,\ V_0 = b$ and

$$\begin{cases} U_{k+1} = & U_k\, V_k \\ V_{k+1} = & V_k \end{cases} \qquad \text{if } \epsilon_k = 0$$

$$\begin{cases} U_{k+1} = & U_k \\ V_{k+1} = & V_k\, U_k \end{cases} \qquad \text{if } \epsilon_k = 1$$

Remark. The set of sequences $\underline{\epsilon}$ satisfying condition (i) is not easy to manipulate as a mathematical object, because in standard topology it is not a compact set.

We know that there is an equivalence between billiard sequences and 2-interval exchanges.

We know too, that induced maps play an essential role and they are related to substitutions on intineraries.

From § 6, we can describe languages L which are factorial and transitive and have a complexity function $p(n) = n + 1$, by a sequence of substitutions.

We are able then to state a fundamental result:

Proposition 2.6.1. Let L be a language.

Suppose that L is factorial, transitive and that is complexity function $p(n)$ verifies

$$p(n) = n + 1$$

Then L is the language associated to a sequence $\underline{y}$ obtained from a square billiard like in § 2.3, and by the proposition we proved in that paragraph, the converse is true i.e. the language associated to any non periodic sequence $\underline{y}$ obtained from a square billiard possesses the properties we quoted above.

A language L such as above is called a Sturmian language.

A sequence $\underline{x}$ such that $L(\underline{x})$ is a Sturmian language is called a Sturmian sequence.

Proposition 3 could also be formulated.

Proposition 2.6.1′. A sequence $\underline{x}$ is a Sturmian sequence, if and only if it can be obtained from a square billiard.

References

[1] Allouche, J.-P., Sur la Complexité des Suites Infinies, *Bull. Belg. Math. Soc.* **1** (1994).

[2] Arnoux, P., G. Rauzy, Representation Geometrique des Suites de Complexité $2n+1$, *Bull. Soc. Math. France* **119** (1991).

[3] Coven, E., G. Hedlund, Sequences with Minimal Blocks Growth, *Math. Systems Theory* **7** (1973).

[4] Hedlund, G., Sturmian Minimal Sets, *Amer. J. Math.* **66** (1944).

[5] Hedlund, G., M. Morse, Symbolic Dynamics II. Sturmian Trajectories, *Amer. J. Math.* **62** (1940).

[6] Hubert, P., Complexité des Suites Definies par des Trajectoires de Billard dans un Polygone Rationnel, *Bull. Soc. Math. France* (1993).

[7] Queffelec, M., Substitution Dynamical Systems - Spectral Analysis, *Lecture Notes in Mathematics* **1294**, Springer, Berlin, (1987).

[8] Rauzy, G., Suites a Termes dans un Alphabet Fini, Seminaire de Theorie des Nombres de Bordeaux (1982-1983).

[9] Zorich, A., Deviation for the Interval Exchange Transformations, Preprint Max-Planck Institue für Mathematik Bonn.

Nonlinear Phenomena and Complex Systems

1. E. Tirapegui and W. Zeller (eds.): *Instabilities and Nonequilibrium Structures V*. 1996 ISBN 0-7923-3992-4
2. E. Goles and S. Martínez (eds.): *Dynamics of Complex Interacting Systems*. 1996 ISBN 0-7923-4173-2

KLUWER ACADEMIC PUBLISHERS – DORDRECHT / BOSTON / LONDON

MIX
Papier aus verantwortungsvollen Quellen
Paper from responsible sources
FSC® C105338

If you have any concerns about our products,
you can contact us on
ProductSafety@springernature.com

In case Publisher is established outside the EU,
the EU authorized representative is:
Springer Nature Customer Service Center GmbH
Europaplatz 3, 69115 Heidelberg, Germany

Printed by Libri Plureos GmbH
in Hamburg, Germany